Luigi Amerio (Ed.)

Equazioni differenziali astratte

Lectures given at the
Centro Internazionale Matematico Estivo (C.I.M.E.),
held in Varenna (Como), Italy,
May 30-June 8, 1963

C.I.M.E. Foundation
c/o Dipartimento di Matematica "U. Dini"
Viale Morgagni n. 67/a
50134 Firenze
Italy
cime@math.unifi.it

ISBN 978-3-642-11003-0 e-ISBN: 978-3-642-11005-4
DOI:10.1007/978-3-642-11005-4
Springer Heidelberg Dordrecht London New York

Reprint of the 1st ed. C.I.M.E., Ed. Cremonese, Roma, 1963
With kind permission of C.I.M.E.

Printed on acid-free paper

Springer.com

CENTRO INTERNATIONALE MATEMATICO ESTIVO
(C.I.M.E)

Reprint of the 1st ed.- Varenna, Italy, May 30-June 8, 1963

EQUAZIONI DIFFERENZIALI ASTRATTE

CENTRO INTERNAZIONALE MATEMATICO ESTIVO

(C. I. M. E.)

T O S I O K A T O

SEMI-GROUPS AND TEMPORALLY INHOMOGENOUS EVOLUTION EQUATIONS

ROMA - Istituto Matematico dell'Università

SEMI-GROUPS AND TEMPORALLY INHOMOGENOUS EVOLUTION EQUATIONS

by T. KATO

INTRODUCTION

These lectures are concerned with the Cauchy problem for the time-indépendent evolution equation

$$\text{(E)} \qquad \frac{du}{dt} + A(t)u = f(t), \qquad 0 < t \leq T; \qquad u(o) = u_o.$$

The unknown $u = u(t)$ and the given function $f(t)$ take values in a Banach space X ; $A(t)$ is a (in general unbounded) linear operator in X depending on t .

It will suffice to mention here only a few examples of (E).

Ex. 1. A parabolic differential equation

$$\frac{\partial u}{\partial t} - \sum_{j,k=1}^{n} a_{jk}(x,t) \frac{\partial^2 u}{\partial x_j \partial x_k} - \sum_{j=1}^{n} a_j(x,t) \frac{\partial u}{\partial x_j} - a(x,t)u = f(x,t)$$

$$x \in \Omega \subset R^n, \qquad 0 < t \leq T$$

is in the form (E) with an obvious definition of A(t). The boundary conditions, which may depend on t , are included in the definition of A(t).

Ex. 2. The Schrödinger equation

$$\frac{1}{i} \frac{\partial u}{\partial t} + \Delta u - V(x,t)u = 0, \qquad x \in R^3$$

also has the form (E). Here $A(t) = i(\Delta - V(.\,,t))$ is i times a self-adjoint operator (at least formally) in $X = L^2(R^3)$.

Ex. 3. The wave equation

$$\frac{\partial^2 u}{\partial t^2} = \Delta u$$

in $x \in R^3$ may be reduced to the form (E) by writing

$$\frac{\partial}{\partial t}\begin{pmatrix} u \\ v_1 \\ v_2 \\ v_3 \end{pmatrix} - \begin{pmatrix} 0 & \frac{\partial}{\partial x_1} & \frac{\partial}{\partial x_2} & \frac{\partial}{\partial x_3} \\ \frac{\partial}{\partial x_1} & 0 & 0 & 0 \\ \frac{\partial}{\partial x_2} & 0 & 0 & 0 \\ \frac{\partial}{\partial x_3} & 0 & 0 & 0 \end{pmatrix}\begin{pmatrix} u \\ v_1 \\ v_2 \\ v_3 \end{pmatrix} = 0$$

where v_1, v_2, v_3 are auxiliary functions. This has the form (E), where u is replaced by the 4-component vector function (u, v_1, v_2, v_3).

In what follows I want to deduce several sufficient conditions on A(t) and f(t) in order that (E) has a unique solution.

§ 1. GENERATION OF DIFFERENT TYPES OF SEMI-GROUPS

1. Let us consider (E) first in the special case when $A(t) = A$ is independent of t :

$$(E_o) \qquad \frac{du}{dt} + Au = f(t) , \qquad u(o) = u_o .$$

The solution is formally given by

$$(S_o) \qquad u(t) = e^{-tA} u_o + \int_o^t e^{-(t-s)A} f(s) ds .$$

The problem is, therefore, essentially that of constructing the exponential function e^{-tA}. This is exactly the problem of generating a semi-group $\{ e^{-tA} \}$ from a given operator A .

In what follows we consider only <u>strongly continuous semi-groups</u> on $[0, \infty)$. Thus $\{ e^{-tA} \}$ is a semi-group if and only if e^{-tA} is strongly continuous for $0 \leq t < \infty$ with $e^{-0A} = 1$ and has the semi-group property : $e^{-tA} e^{-sA} = e^{-(t+s)A}$ for all s, $t \geq 0$.

The generation of such semi-groups has been discussed in detail in Professor Phillips' lectures. We reproduce here some of his results that we need, with several additional remarks.

<u>Definition 1. 1.</u> We say $-A \in (Bo)$ if

1) A is densely defined and closed.

2) any $\lambda > o$ belongs to the resolvent set $p(-A)$ of $-A$, with

$$| (\lambda +A)^{-n} | \leq \frac{M}{\lambda^n} , \qquad n = 1, 2, 3, \ldots .$$

where M is a constant independent of λ or n .

<u>Theorem 1. 2.</u> Let $-A \in (Bo)$. Then there exists a unique semi-group $\{ e^{-tA} \}$ with $| e^{-tA} | \leq M$ such that

$$(D) \qquad \frac{d}{dt} e^{-tA} u = -Ae^{-tA} u = - e^{-tA} Au$$

for $u \in D_{A'}$ e^{-tA} commutes with $(\lambda + A)^{-1}$.

Proof. See Phillips' lectures

Remark 1. 3. If $-A \in (Bo)$, it follows that all complex λ with $\operatorname{Re} \lambda > 0$ belong to $\rho(-A)$, with

$$\left|(\lambda + A)^{-n}\right| \leq \frac{M}{(\operatorname{Re} \lambda)^n} \qquad n = 1, 2, \ldots .$$

This is seen by considering the Laplace transforms of $t^{n-1} \cdot e^{-tA}$ (see Phillips).

Definition 1. 4. If $M - 1$ above, $\{e^{-tA}\}$ is called a contraction semi-group. The subset of (Bo) determined by $M = 1$ will be denoted by (Co). (Note that $M \geq 1$ in general).

2. We now introduce another subset of (Bo) which is important in our problems.

Definition 1. 5. We say $-A \in (Ho)$ if

1) A is densely defined and closed

2) The spectrum $\sigma(A)$ of A is a subset of a sector

$|\arg \lambda| \leq \frac{\pi}{2} - \omega$, $\omega > 0$, and

$$\left|(\lambda + A)^{-1}\right| \leq \frac{M_\varepsilon}{|\lambda|} \qquad \text{for} \qquad |\arg \lambda| \leq \frac{\pi}{2} + \omega - \varepsilon$$

Remark 1. 6. (Ho) $\subset$ (Bo). This is not obvious from the definition, but follows from Theorem 1. 7. below (again take the Laplace transforms of $t^{n-1} e^{-tA}$).

Theorem 1. 7. Let $-A \in (Ho)$. Then there exists a unique semi-group $\{e^{-tA}\}$ such that for $t > 0$

$$\text{(D')} \qquad -e^{-tA} A \supset \frac{d}{dt} e^{-tA} = -A e^{-tA} \in B^{1)}.$$

---

(1) We denote by B the set of all bounded linear operator in X with domain X .

$\{e^{-tA}\}$ can be continued analytically to the sector $|\arg t| < \omega$, $t \neq 0$, with (D') preserved. Furthermore, e^{-tA} and tAe^{-tA} are uniformly bounded in any smaller sector :

$$|e^{-tA}| \leq M'_{\varepsilon}, \quad |t|\,|Ae^{-tA}| \leq M'_{\varepsilon}, \quad |\arg t| \leq \omega - \varepsilon,$$

and $e^{-tA} \longrightarrow 1$ strongly when $t \longrightarrow 0$ in this smaller sector [1].

Proof. We can define e^{-tA} by the Dunford integral :

$$(1) \qquad e^{-tA} = \frac{1}{2\pi i} \int_C e^{\lambda t} (\lambda + A)^{-1} d\lambda \in \mathcal{B}, \quad t > 0,$$

where C is a curve, running in $\rho(-A)$, from $\infty e^{-i\vartheta}$ to $\infty e^{i\vartheta}$ where $\frac{\pi}{2} < \vartheta < \frac{\pi}{2} + \omega$. Thus the integral is absolutely convergent and defines an operator of $\mathcal{B}$.

The semi-group property follows from the standard argument with Dunford integrals. We have namely

$$(2) \qquad e^{-sA} = \frac{1}{2\pi i} \int_{C'} e^{\lambda' s} (\lambda' + A)^{-1} d\lambda',$$

where C' is obtained from C by a slight shift to right. Multiplying (1), (2) and using the resolved equation, we have

$$e^{-tA} e^{-sA} = \left(\frac{1}{2\pi i}\right)^2 \int_C \int_{C'} e^{\lambda t + \lambda' s} \frac{1}{\lambda' - \lambda} \left[(\lambda + A)^{-1} - (\lambda' + A)^{-1}\right] d\lambda\, d\lambda',$$

where the order of integration is arbitrary.

$$\text{Now} \quad \int_C e^{\lambda t} \frac{d\lambda}{\lambda' - \lambda} = 0 \quad \text{and} \quad \int_{C'} \frac{e^{\lambda' s}}{\lambda' - \lambda} d\lambda' = 2\pi i\, e^{\lambda s}$$

since C lies to left of C'. Hence

$$e^{-tA} e^{-sA} = \frac{1}{2\pi i} \int_C e^{\lambda (t+s)} (\lambda + A)^{-1} d\lambda = e^{-(t+s)A},$$

---

(1) For analytic semi-groups, see [15] as well as the book by Hille-Phillips.

proving the semi-group property of $\{e^{-tA}\}$.

That e^{-tA} has an analytic continuation is obvious from (1). In fact the integral of (1) converges for any t with $|\arg t| < \vartheta - \frac{\pi}{2}$, hence for any t with $|\arg t| < \omega$, by taking ϑ suitably. Moreover,

$$\text{(3)} \qquad \frac{d}{dt} e^{-tA} = \frac{1}{2\pi i} \int_C \lambda e^{\lambda t} (\lambda + A)^{-1} d\lambda \in B, \qquad t \neq 0.$$

Since $\lambda(\lambda + A)^{-1} = 1 - A(\lambda + A)^{-1}$ and $\int_C \lambda e^{\lambda t} d\lambda = 0$, (D') follows from (3) (note that $A(\lambda + A)^{-1} A \subset A(\lambda + A)^{-1}$).

To prove the uniform boundedness of e^{-tA}, we change the integration variable from λ to $\lambda' = \lambda t$ in (1). The corresponding integration path tC can be deformed to a path C', independent of t, which runs from $\infty e^{-i\vartheta'}$ to $\infty e^{ti\vartheta'}$ with $\vartheta' = \frac{\pi}{2} + \varepsilon$, $\varepsilon > 0$ being very small.

The resulting expression

$$e^{-tA} = \frac{1}{2\pi i} \int_{C'} e^{\lambda'} \left(\frac{\lambda'}{t} + A\right)^{-1} \frac{d\lambda'}{t}$$

is true for any $t \neq 0$ with $|\arg t| \leq \omega - \varepsilon$. Since

$\left|\left(\frac{\lambda'}{t} + A\right)^{-1}\right| \leq M / \left|\frac{\lambda'}{t}\right| = M|t| / |\lambda'|$, it follows that

$$\text{(3)'} \qquad |e^{-tA}| \leq \frac{M}{2\pi} \int_{C'} |e^{\lambda'}| \frac{|d\lambda'|}{|\lambda'|} = M'_{\varepsilon} .$$

In the same way one proves $|Ae^{-tA}| \leq M''_{\varepsilon} / |t|$.

To prove $e^{-tA} \to 1$, $t \to 0$, we note that

$$(e^{-tA} - 1)u = \frac{1}{2\pi i} \int_C e^{\lambda t} [(\lambda + A)^{-1} - \lambda^{-1}] u \, d\lambda = \frac{-1}{2\pi i} \int e^{\lambda t} (\lambda + A)^{-1} Au \frac{d\lambda}{\lambda}$$

if $u \in D_A$. Hence for $t \to 0$

$$(e^{-tA} - 1)u \to -\frac{1}{2\pi i} \int (\lambda + A)^{-1} Au \frac{d\lambda}{\lambda} = 0$$

(the integrand is $O(\lambda^{-2})$ for $\lambda \to \infty$, $\mathrm{Re}\,\lambda > 0$).

Since e^{-tA} is uniformly bounded for $|\arg t| \leq \omega - \varepsilon$ as proved above and since D_A is dense, this proves that $e^{-tA} \to 1$ strongly, q. e. d. [1].

Remark 1. 8. Theorem 1. 7. implies that, if $-A \in (H0)$, e^{-tA} sends X into D_A and $e^{-tA}u$ is always differentiable for any $u \in X$, if $t \neq 0$. This is a great difference from the case $-A \in (B0)$, where $e^{-tA}u \in D_A$ is in general expected only for $u \in D_A$.

Remark 1. 9. There are many examples of operators of (H0). Generally speaking, any strongly elliptic partial differential operator with "ordinary" boundary conditions belongs to (H0). Furthermore if X is a Hilbert space, there is a rather general sufficient condition for $-A \in (H0)$. Suppose that the numerical range $N_A = \{(Au, u) \,|\, |u| = 1, \ u \in D_A\}$ of A is a subset of a sector $|\arg \lambda| \leq \frac{\pi}{2} - \omega$, $\omega > 0$. If, in addition, there is at least one point λ exterior to N_A that belongs to $\rho(A)$, then $-A \in (H0)$[2].

3. We now consider the solution of the inhomogeneous equation (E_0).

Definition 1. 10. By a solution of (E) we mean a function $u(t)$ with the following properties.

1) $u(t)$ is (strongly) continuous for $0 \leq t \leq T$, $u(o) = u_0$.

2) $u(t)$ is (strongly) differentiable for $0 < t \leq T$.

3) $u(t) \in D_{A(t)}$ for $0 < t \leq T$ so that $A(t)u(t)$ makes sense.

4) (E) is true for $0 < t \leq T$.

The same definition applies to (E_0) when $A(t) = A$ is constant.

Theorem 1. 11. Let $-A \in (B0)$. Then any solution of (E_0) is given by (S_0) if $f(t)$ is continuous for $0 \leq t \leq T$. Conversely, $u(t)$ given by (S_0) is a solution of (E_0) if $u_0 \in D_A$ and $f(t)$ is continuously differentiable. In this ca-

---

(1) The uniqueness of $\{e^{-tA}\}$ with the properties stated follows from Theorem 1. 2.

(2) This is due to the fact that $|(\lambda + A)u| \geq d_\lambda |u|$ for any $u \in D_A$ where d_λ is the distance of λ from N_{-A}. It follows, under the condition stated, that $|(\lambda + A)^{-1}| \leq 1/d_\lambda \leq M/|\lambda|$.

se $Au(t)$ and $du(t)/dt$ are continuous [1].

Proof. Let $u(t)$ be a solution of (E_o). Then

$$\frac{d}{ds} e^{-(t-s)A} u(s) = e^{-(t-s)A} Au(s) + e^{-(t-s)A}(-Au(s)+f(s)) =$$
$$= e^{-(t-s)} f(s)$$

since $u(s) \in D_A$ and $e^{-(t-s)A} \in B$. Integration on s then gives (S_o) immediately.

Conversely, suppose $u_o \in D_A$ and $f(t)$ is continuously differentiable. Since $e^{-tA}u_o$ satisfies the homogeneous equation and the initial condition by Theorem 1.2, we need only to consider the second term of (S_o). In other words, we may assume $u_o = 0$.

Noting that $f(s) = f(o) + \int_o^s f'(r)dr$, we have then

$$u(t) = \int_o^t e^{-(t-s)A} f(o)ds + \int_o^t dr \int_r^t e^{-(t-s)A} f'(r)ds .$$

But (see Lemma 1.12 below)

$$A\int_o^t e^{-(t-s)A} ds = A\int_o^t e^{-sA} ds = 1 - e^{-tA} ,$$

$$A\int_r^t e^{-(t-s)A} ds = A\int_o^{t-r} e^{-sA} ds = 1 - e^{-(t-r)A} .$$

Hence $Au(t)$ exists and

$$Au(t) = (1 - e^{-tA})f(o) + \int_o^t (1 - e^{-(t-r)A})f'(r)dr$$
$$= f(t) - e^{-tA}f(o) - \int_o^t e^{-sA} f'(t-s)ds .$$

On the other hand

(1) This theorem is due to Phillips [8] .

$$\frac{d}{dt} u(t) = \frac{d}{dt} \int_0^t e^{-sA} f(t-s)ds = e^{-tA} f(o) + \int_0^t e^{-sA} f'(t-s)ds$$

Hence $\frac{d}{dt} u(t) = - Au(t) + f(t)$, as we wished to show.

By the way, the continuity of $\frac{d}{dt} u(t)$ and of $Au(t)$ are obvious from the above expressions.

Lemma 1. 12. Let $-A \in (Bo)$. Then

$$A \int_r^t e^{-sA} ds = e^{-rA} - e^{-sA} , \qquad 0 \leq r < t .$$

Proof. If $u \in D_A$, we have $Ae^{-sA}u = - \frac{d}{ds} e^{-sA}u = e^{-sA}Au$ (which is continuous in t).

Hence

$$A \int_r^t e^{-sA} u \, ds = \int_r^t Ae^{-sA} u \, ds = (e^{-rA} - e^{-tA})u .$$

(The first equality is a direct consequence of the closure of A). For any $v \in X$, let $u_n \in D_A$, $u_n \to v$. Then $\int_r^t e^{-sA} u_n ds \to \int_r^t e^{-sA} v \, ds$ and $A(\int_r^t e^{-sA} u_n ds) = (e^{-rA} - e^{-tA}) u_n \to (e^{-rA} - e^{-tA}) v$. It follows, again by the closure of A, that $A \int_r^t e^{-sA} v \, ds$ exists and equals $(e^{-rA} - e^{-tA}) v$.

q. e. d.

Theorem 1. 13. If $-A \in (H0)$, the continuous differentiability of $f(t)$ in the second part of Theorem 1. 11 can be replaced by a Hölder continuity. Furthermore, $u(t)$ of (S_o) is analytic if $f(t)$ is analytic on $[0, T]$.

Proof. Again we may assume $u_o = 0$. Then

$$u(t) = \int_0^t e^{-(t-s)A} (f(s) - f(t))ds + \int_0^t e^{-(t-s)A} f(t)dt .$$

Therefore (see Lemma 1. 12)

$$Au(t) = \int_0^t A \, e^{-(t-s)A} (f(s) - f(t))ds + (1 - e^{-tA}) f(t) .$$

Note that the integral exists because $\left| Ae^{-(t-s)A} \right| \leq \frac{\text{const}}{t-s}$ (see Theorem 1. 7) and $|f(s) - f(t)| \leq \text{const}(t-s)^{\vartheta}$, $\vartheta > 0$.

This shows that $Au(t)$ exists and (closure of A(t)!)

$$A(t)u = A \int_0^t e^{-(t-s)A}(f(s) - f(t))\, ds + (1 - e^{-tA})f(t) .$$

On the other hand, the construction of $\partial u(t)/\partial t$ requires a little detour. We define

$$u_{\varepsilon}(t) = \int_0^{t-\varepsilon} e^{-(t-s)A} f(s)\, ds$$

Obviously $u_{\varepsilon}(t) \longrightarrow u(t)$ for $\varepsilon \to 0$, locally uniformly in t . Also

$$\frac{d}{dt} u_{\varepsilon}(t) = e^{-\varepsilon A} f(t-\varepsilon) - \int_0^{t-\varepsilon} A\, e^{-(t-s)A} f(s)\, ds \ ;$$

the integral exists since $Ae^{-(t-s)A}$ is continuous for $s \leq t - \varepsilon$. But it is easy to see that the limit for $\varepsilon \to 0$ of this integral exists and equals $Au(t)$ (use again the Hölder continuity of $f(t)$), so that

$$\frac{d}{dt} u_{\varepsilon}(t) \to f(t) - Au(t) .$$

Moreover, this convergence is locally uniform in t . Hence it follows that $\frac{d}{dt} u(t)$ exists and equals $f(t) - Au(t)$, by a well known theorem in differential calculus.

If $f(t)$ is analytic, $u_{\varepsilon}(t)$ is also analytic : $du_{\varepsilon}(t)/dt$ given above exists for complex t in some neighborhood of the interval $[2\varepsilon , T]$. But $u_{\varepsilon}(t) \to u(t)$ is true locally uniformly in t for these complex t . It follows that $u(t)$ is analytic. q. e. d.

§ 2. THE CASE IN WHICH -A(t) ARE GENERATORS OF ANALYTIC SEMI-GROUPS WITH CONSTANT DOMAIN FOR $A(t)^h$.

1. First we note that the equation (E) is very simply dealt with if $A(t) \in B$ and strongly continuous in t. If we consider the homogeneous equation $du/dt + A(t)u = 0$, the solution can be constructed by a straightforward successive approximation :

$$u(t) = \sum_{k=0}^{\infty} u_k(t) \;, \quad u_0(t) = u_0 \;,$$

$$u_k(t) = -\int_0^t A(s)u_{k-1}(s)\, ds \;, \qquad k = 1, 2, 3, \ldots$$

This is equivalent to writing $u(t) = U(t,0)u_0$ and determining $U(t,0)$ from the differential equation $dU(t,0)/dt = -A(t)U(t,0)$, $U(0,0) = 1$, by successive approximation (the derivative is strong derivative). More generally, we can solve the differential equation

$$\text{(1)} \qquad \frac{\partial}{\partial t} U(t,s) = -A(t)U(t,s), \qquad U(s,s) = 1$$

by successive approximation. The family of operators $U(t,s)$ constructed in this way will be called the evolution operator (or the Green function).

The evolution operator has, in addition to (1), the following properties:

$$\text{(2)} \qquad \frac{\partial}{\partial s} U(t,s) = U(t,s)\, A(s)$$

$$\text{(3)} \qquad U(t,s)\, U(s,r) = U(t,r) \;.$$

To prove this, it is convenient to consider another differentiable equation

$$\frac{\partial}{\partial s} V(t,s) = V(t,s)\, A(s), \qquad V(t,t) = 1 \;.$$

This can again be solved by successive approximation.

Then

$$\frac{\partial}{\partial s} V(t,s)\,U(s,r) = V(t,s)(A(s) - A(s))\,U(s,r) = 0$$

so that $V(t,s)\,U(s,r)$ is independent of s. Putting $s = t$ and $s = r$, we obtain $U(t,r) = V(t,r)$, and hence $U(t,s)\,U(s,r) = U(t,r)$, q. e. d.

With the use of the evolution operator, the solution of (E) can be expressed by

(S) $$u(t) = U(t,0)u_0 + \int_0^t U(t,s)\,f(s)\,ds \,.$$

Now the above method does not work when $A(t)$ is not bounded. Therefore we want to construct the evolution operator for unbounded $A(t)$ by a limiting procedure, by approximating $A(t)$ by a sequence $A_n(t)$ of bounded operators (this is the way the semi-group e^{-tA} was constructed in Phillips' lectures as the limit of e^{-tA_n}, A_n being bounded). We choose

(4) $$A_n(t) = A(t)J_n(t) = n(1 - J_n(t)) \,, \quad J_n(t) = (1 + \frac{1}{n} A_n(t))^{-1} \,, \quad n = 1, 2, 3, ..$$

If $A_n(t)$ (or $J_n(t)$) is strongly continuous in t, we can construct the evolution operator $U_n(t,s)$ by the simple method described above. Then we want to show that $\text{s-lim}\, U_n(t,s)$ exists, which will be the evolution operator $U(t,s)$ for the unbounded case.

This method is seen to work under certain conditions on $A(t)$. We have namely

<u>Theorem 2. 1.</u>[1)] Assume that

i) $-A(t) \in (H0)$ uniformly for $0 \leq t \leq T$. This means that there are constants $M > 0$, $\omega > 0$ such that

---

(1) This theorem is first proved by Tanabe (see [10, 11, 12]). The last proposition regarding the analyticity is due to Komatsu [7]. The proof given here is somewhat different from theirs. See also Yosida [17].

$$|(\lambda + A(t))^{-1}| \leq \frac{M}{|\lambda|} \quad \text{for} \quad |\arg\lambda| \leq \frac{\pi}{2} + \omega$$

and

$$|A(t)^{-1}| \leq M .$$

ii) $D_{A(t)} = D$ is independent of t. This implies that $A(t)A(s)^{-1} \in B$ for any s and t.

iii) $A(t)\,A(0)^{-1}$ is Hölder continuous (in norm). This is equivalent to that

$$|A(t)\,A(s)^{-1} - 1| \leq M(t-s)^{\vartheta} , \qquad \vartheta > 0 .$$

Then there exist a unique evolution operator $U(t,s)$ with the following properties :

1) $U(t,s) \in B$ and strongly continuous for $0 \leq s \leq t \leq T$.

2) $U(t,s)\,U(s,r) = U(t,r)$, $U(t,t) = 1$.

3) $U(t,s)$ is strongly continuously differentiable in t for $t > s$, with

$$\frac{\partial}{\partial t} U(t,s) = - A(t)\,U(t,s) , \qquad |A(t)\,U(t,s)| \leq \frac{\text{const.}}{t-s} .$$

4) If $u \in D$, then

$$\frac{\partial}{\partial s} U(t,s)\,u = U(t,s)\,A(s)\,u , \qquad s \leq t .$$

If, in addition, $A(t)^{-1}$ is analytic in t , then $U(t,s)$ has an analytic continuation for complex s,t such that $|\arg(t-s)|$ are sufficiently small.

Remark 2. 2. a) ii) and iii) express that $A(t)$ depends on t "smoothly". Note that ω of ii) is equal to the $\omega - \varepsilon$ of Def. 1. 5. for some fixed ε .
b) The assumption $|A(t)^{-1}| \leq M$ is made only for simplicity; if this is not the case, then we may make a transformation $u(t) \longrightarrow e^{-\varepsilon t}u(t)$ in (E) so that the new equation has $A(t)$ satisfying the above condition (this is allowed since we are interested only in a finite interval $0 \leq t \leq T$). c) The assumptions ii) and iii) will be weakened later. But it should be remarked that these

are satisfied if $A(t)$ is an operator in $X = L^2(\Omega)$ determined from an elliptic differential operator with smooth coefficients on a bounded region Ω with a smooth boundary, with the Dirichlet boundary condition. In this case $D_A(t) = H^{2m}(\Omega) \cap H_0^m(\Omega)$ and is independent of t.

2. Proof of Theorem 2. 1.

We construct the approximating operators $A_n(t)$ by (4). $A_n(t)$ is bounded by $|A_n(t)| \le (M+1)n$ since $|J_n(t)| \le M$. But the important fact is that $-A_n(t)$ belongs to the class (H0) uniformly in t and n in the sense stated in i) of Theorem 2. 1'. We have namely

(5) $$A_n(t)^{-1} = A(t)^{-1} + n^{-1}, \qquad |A_n(t)^{-1}| \le M + n^{-1} \le M+1.$$

Also a straightforward computation leads to

(6) $$(\lambda + A_n(t))^{-1} = \frac{n^2}{(n+\lambda)^2}\left(\frac{n\lambda}{n+\lambda} + A(t)\right)^{-1} + \frac{1}{n+\lambda}$$

from which an elementary geometric consideration gives

(7) $$|(\lambda + A_n(t))^{-1}| \le \frac{M'}{|\lambda|} \qquad |\arg\lambda| \le \frac{\pi}{2} + \omega$$

where M' is in general different from M but may be taken independent of n. Finally, we have

(7') $$A_n(t)\,A_n(s)^{-1} = A_n(t)\,(A(s)^{-1} + \frac{1}{n}) = J_n(t)\,A(t)\,A(s)^{-1} + 1 - J_n(t),$$
$$A_n(t)\,A_n(s)^{-1} - 1 = J_n(t)\,(A(t)A(s)^{-1} - 1),$$

hence

(8) $$|A_n(t)\,A_n(s)^{-1} - 1| \le M\,|A(t)A(s)^{-1} - 1| \le M^2(t-s)^{\varrho}$$

In particular $|A_n(t) - A_n(s)| \le |A_n(t)\,A_n(s)^{-1} - 1|\,|A_n(s)| \le$ $\le M^2(M+1)\,n(t-s)^{\varrho}$, so that $A_n(t)$ is itself Hölder continuous.

Therefore, the evolution operator $U_n(t,s)$ for $A_n(t)$ can be constructed as stated above.

To prove that $\text{s-lim}_{n\to\infty} U_n(t,s)$ exists, however, we have to deduce other expressions and estimates for $U_n(t,s)$. To this end we first note the identity

$$\frac{\partial}{\partial s} U_n(t,s)e^{-(s-r)A_n(r)} = U_n(t,s)(A_n(s) - A_n(r))e^{-(s-r)A_n(r)}$$

Integrating on $r \leq s \leq t$, one obtains

$$U_n(b,r) = e^{-(t-r)A_n(r)} + \int_r^t U_n(t,s)\, K_n(s,r)ds\,, \tag{9}$$

where

$$K_n(s,r) = -\left[A_n(s)\, A_n(r)^{-1} - 1\right] A_n(r)e^{-(s-r)A_n(r)} \quad . \tag{10}$$

$K_n(s,r)$ has the estimate

$$\left| K_n(s,r) \right| \leq \frac{C}{(s-r)^{1-\theta}} \quad , \tag{11}$$

where C is a constant independent of n. This follows from (8) and $\left| A_n(r)e^{-(s-r)A_n(r)} \right| \leq C(s-r)^{-1}$, which is in turn a consequence of $-A_n(r) \in (H0)$ (Theorem 1. 7.).

(9) may be considered an integral equation for $U_n(t,r)$. This may be written symoblically

$$U_n = U_n^{(o)} + U_n * K_n \tag{12}$$

and can be solved by successive approximation :

$$U_n = \sum_{k=o}^{\infty} U_n^{(k)} \quad , \qquad U_n^{(k)} = U_n^{(k-1)} * K_n \quad . \tag{13}$$

The possibility of this successive approximation is guaranteed by the estimate (11) [1)] for K_n and the estimate $\left|U^{(o)}(t,r)\right| \leq C$ (we use the same symbol

---

(1) The important fact is that K_n are dominated by an integrable kernel independent of n.

C to denote different constants). It follows that the series (13) is convergent uniformly, being majorized by a series <u>independent of n</u> .

Now we can let $n \to \infty$. Then

(14) $$U_n^{(o)}(t,s) \underset{s}{\longrightarrow} U^{(o)}(t,s) = e^{-(t-s)A(s)} .$$

(14) is exactly the construction of the semi-group $e^{-\tau A(s)}$ as the limit of $e^{-\tau A_n(s)}$ (see Phillips' lectures).

Also

(15) $$K_n(t,s) \underset{s}{\longrightarrow} K(t,s) = -(A(t)A(s)^{-1}-1)A(s)e^{-(t-s)A(s)}$$

by (7') and $J_n(t) \longrightarrow 1$ (which is a basic fact proved in the generation of semi-groups, see Phillips) and

(16) $$A_n(s)e^{-\tau A_n(s)} \underset{s}{\dashrightarrow} A(s)e^{-\tau A(s)} , \qquad \tau > 0 ,$$

which follows from (3) of § 1. We note that the strong convergence (14) and (15) are uniform in s and t (for $t-s \geq \alpha > 0$ in (15)).

It follows from (14) and (15) that $U_n^{(1)}(t,s) = U_n^{(o)} * K_n(t,s) \underset{s}{\longrightarrow} U^{(1)}(t,s)$, then $U_n^{(2)}(t,s) = U_n^{(1)} * K_n(t,s) \underset{s}{\longrightarrow} U^{(2)}(t,s)$, and so on. In view of the fact that the series (13) is uniformly majorized, we conclude that

(17) $$U_n(t,s) = \sum_k U_n^{(k)}(t,s) \underset{s}{\longrightarrow} U^{(k)}(t,s) \equiv U(t,s), \quad |U(t,s)| \leq C .$$

Since it is easily seen that the strong convergence (17) is uniform in s, t for $s \leq t$, U(t, s) is strongly continuous for $s \leq t$. Also 2) of Theorem 2. 1. follows from the corresponding relation for $U_n(t,s)$.

3. Proof of Theorem 2. 1., continued.

We use another identity

$$\frac{d}{ds} e^{-(t-s)A_n(t)} U_n(s,r) = e^{-(t-s)A_n(t)}(A_n(t) - A_n(s)) U_n(s,r) ,$$

whence we obtain

$$U_n(t,r) = e^{-(t-s)A_n(t)} + \int_r^t e^{-(t-s)A_n(t)}(A_n(t) A_n(s)^{-1} - 1)A_n(s) U_n(s,r)ds.$$

Multiply this equation with $A_n(t)$ from left and write $Y_n(t,s) = A_n(t) U_n(t,s)$; then

$$Y_n(t,r) = Y_n^{(o)}(t,r) + \int_r^t H_n(t,s) Y_n(s,r)ds \tag{18}$$

where

$$\begin{cases} Y_n^{(o)}(t,s) = A_n(t)e^{-(t-s)A_n(t)} \\ \\ H_n(t,s) = A_n(t)e^{-(t-s)A_n(t)}(A_n(t)A_n(s)^{-1} - 1). \end{cases} \tag{19}$$

(18) may be written symbolically as

$$Y_n = Y_n^{(o)} + H_n * Y_n . \tag{20}$$

We want to solve (20) again by successive approximation :

$$Y_n = \sum_{k=o}^{\infty} Y_n^{(k)} , \qquad Y_n^{(k)} = H_n * Y_n^{(k-1)} . \tag{21}$$

Here, however, we have a slight difficulty that did not exist in (13), for $Y_n^{(o)}$ has the uniform (independent of n) estimate $|Y_n^{(o)}(t,s)| \leq C(t-s)^{-1}$ where $(t-s)^{-1}$ is not integrable. Thus the uniform estimate of $Y_n^{(1)}$ is not quite simple, although its existence is obvious ($A_n(t) \in B$!).

Here we give only the result :

$$|Y_n^{(1)}(t,s)| \leq \frac{C}{(t-s)^{1-\theta}} \tag{22}$$

of which the proof will be given in n. 7.

Once (22) is established, the further successive approximation proceeds smoothly, for the right member of (22) is integrable as well as that of

(23) $$| H_n(t,s) | \leq \frac{C}{(t-s)^{1-\theta}}$$

which is proved as in (11). By an argument similar to that given in n. 2, it follows that (21) is uniformly majorized and that each term $Y_n^{(k)}$ has a strong limit $Y^{(k)}$ for $n \to \infty$. Hence

(24) $$A_n(t)\, U_n(t,s) = Y_n(t,s) \xrightarrow[s]{} Y(t,s), \qquad |Y(t,s)| \leq \frac{C}{t-s}$$

(24) gives (see also (5))

$$U_n(t,s) = A_n(t)^{-1}\, Y_n(t,s) \xrightarrow[s]{} A(t)^{-1}\, Y(t,s)\,.$$

Since $U_n(t,s) \xrightarrow[s]{} U(t,s)$ by (17), we must have $U(t,s) = A(t)^{-1} Y(t,s)$. This means that $A(t)U(t,s)$ exists and equals $Y(t,s) \in B$ if $t > s$. Thus we have proved

(25) $$| A(t)\, U(t,s) | \leq \frac{C}{t-s}$$

The differentiability of $U(t,s)$ is proved in the following way. Since $\partial U_n(t,s)u/\partial t = -A_n(t)U_n(t,s)u$ and $U_n(t,s)u \to U(t,s)u$, $A_n(t)U_n(t,s)u \to$ $\to Y(t,s)u = A(t)U(t,s)u$ uniformly for $t \geq s+a$, it follows that $\partial U(t,s)u/\partial t$ exists and is equal to $-A(t)U(t,s)u$. That is, $U(t,s)$ is strongly differentiable in t for $t > s$, with the strong derivative $-A(t)U(t,s) \in B$.

Similarly, we have

$$\frac{\partial}{\partial s} U_n(t,s)u = U_n(t,s)A_n(s)u \quad .$$

If $u \in D = D_{A(s)}$, we have $A_n(s)u \to A(s)u$, $n \to \infty$, uniformly in s (see Phillips) so that $\frac{\partial}{\partial s} U_n(t,s)u \to U(t,s)A(s)u$. The same argument as above then proves that $\frac{\partial}{\partial s} U(t,s)u = U(t,s)A(s)u$.

If $A(t)^{-1}$ is analytic in t in a neighborhood Δ of $0 \leq t \leq T$, the same is true with $A_n(t) = (A(t)^{-1} + n^{-1})^{-1}$. Therefore $U_n(t,s)$ can be continued analytically to $t \in \Delta$, $s \in \Delta$. Now the expression of $U_n(t,s)$ by the series (13) holds true when the variables t, s, r are supposed to lie on a straight line in Δ having a small angle ϑ with the positive real axis, uniformly with respect to ϑ , and each term $U_n^{(k)}$ is seen to converge for $n \to \infty$ to $U^{(k)}$ uniformly (on the line as well as in ϑ). Thus $U_n(t,s)$ converges strongly and (locally) uniformly to a $U(t,s)$ as long as $|\arg(t-s)|$ are sufficiently small. It follows that $U(t,s)$ is <u>strongly</u> analytic in such a region of t and s . But since strongly analyticity is equivalent to analyticity (in norm), $U(t,s)$ is analytic. This completes the proof of Theorem 2. 1.

4. We now consider the inhomogeneous equation (E).

<u>Theorem 2. 3.</u> Let the assumptions of Theorem 2. 1. be satisfied. Then the conclusions of Theorem 1. 13. are true (with (E_0) and (S_0) replaced by (E) and (S), respectively).

<u>Proof.</u> Almost the same as for Theorem 1. 13. The only modification required is to note that

$$A(t)U(t,s) = A(t)e^{-(t-s)A(t)} + Y'(t,s)$$

$$Y'(t,s) = \sum_{k=1}^{\infty} Y^{(k)}(t,s) \ , \qquad |Y'(t,s)| \leq \frac{C}{(t-s)^{1-\vartheta}}$$

see (20), (22), (24). Hence

$$A(t)\int_r^t U(t,s)ds = \int_r^t A(t)U(t,s)ds = 1 - e^{-(t-r)A(t)} + \int_r^t Y'(t,s)ds$$

by Lemma 1. 12 ($Y'(t,s)$ is absolutely integrable).

5. Generalizations.

To improve Theorems 2. 1. and 2. 3. , we need the fractional powers $A(t)^{\alpha}$ of $A(t)$.

When $-A \in (B0)$, the fractional powers A^{α} can be defined in a natural way [1)]. Here we assume, for simplicity, that $A^{-1} \in B$ in addition. Then we can first construct the Dunford integral

$$(26) \qquad A^{-\alpha} = -\frac{1}{2\pi i}\int_L z^{-\alpha}(z-A)^{-1}\,dz \in B, \qquad \alpha > 0.$$

where L is a curve from $-\infty$ to $-\infty$ passing between $z = 0$ and $\sigma(A)$. The integral converges absolutely since $|(z-A)^{-1}| \leq M/\mathrm{Im}(-z)$. Since this is a Dunford integral it is easy to see that $A^{\alpha+\beta} = A^{\alpha}A^{\beta}$. If $0 < \alpha < 1$, L may be taken as the double ray $(0, -\infty)$, yielding

$$(27) \qquad A^{-\alpha} = \frac{\sin\pi\alpha}{\pi}\int_0^{\infty}\lambda^{-\alpha}(\lambda+A)^{-1}\,d\lambda .$$

Then we define A^{α} as the inverse of $A^{-\alpha}$; note that $A^{-\alpha}$ is invertible since $A \;\; u = 0$ implies $A^{-n}u = A^{-(n-\alpha)}A^{-\alpha}u = A^{-n}u = 0$, $u = 0$, where n is a positive integer larger than α.

We need also the following expression, which is valid for $-A \in (H0)$.

$$(27') \qquad A^{\alpha}e^{-\tau A} = \frac{1}{2\pi i}\int_C (-\lambda)^{\alpha}e^{\lambda\tau}(\lambda+A)^{-1}\,d\lambda .$$

This can be proved by verifying that (27') gives $e^{-\tau A}$ when multiplied by (26) (cf. the proof of Theorem 1. 7). It follows from (27') that

$$(27'') \qquad |A^{\alpha}e^{-\tau\alpha}| \leq \frac{C}{|\tau|^{\alpha}}$$

---

(1) These fractional powers are considered by many authors; see, for example, [3], [4], [16] and the references given there.

We can now state generalization of Theorems 2. 1. and 2. 3.

Theorem 2. 4.[1)] Assume that

i) $-A(t) \in (H0)$ uniformly (as in i), Theorem 2. 1.).

ii) $D_{A(t)^h} = D_h$ = const. for some $h = 1/m$ with a positive integer m. This implies that $A(t)^h A(s)^{-h} \in B$ for any s and t.

iii) $A(t)^h A(\varrho)^{-h}$ is Hölder continuous with an exponent $\varrho > 1 - h$, so that $| A(t)^h A(s)^{-h} - 1 | \leq M(t-s)^{\varrho}$.

Then the conclusions of Theorem 2. 1. are true (with D replaced by D_h in 4)). In the last statement of Theorem 2. 1. (analyticity), the analyticity of $A(t)^{-h}$ should be assumed.

Theorem 2. 5. Under the assumptions of Theorem 2. 4., the conclusions of Theorem 2. 3. are true.

Remark 2. 6. Theorem 2. 1. is a special case of Theorem 2. 4. for $m = 1$. The assumption that $D_{A(t)^h}$ = const. is supposed to be weaker than that $D_{A(t)}$ = const., but there is no general proof valid for Banach spaces X. In any case this is true for accretive operators $A(t)$ in a Hilbert space. In other words, $D_A = D_B$ implies $D_{A^\alpha} = D_{B^\alpha}$ if X is a Hilbert space and $-A, -B \in (B0)$ are such that $\mathrm{Re}(Au, u) \geq 0$, $\mathrm{Re}(Bu, u) \geq 0$ $(u \in D_A = D_B)$. Furthermore, it has been proved by Lions that, when A is an operator in $X = L^2(\Omega)$ determined from a strongly elliptic differential operator of order $2m$ on a domain Ω with a smooth boundary, A^α has a domain independent of the coefficients or of boundary conditions for $\alpha < 1/4m$.

Outline of the proof. Here it is impossible to give a complete proof which is in principle the same as the proof of Theorem 2. 1. We shall indicate briefly the essential points in the proof.

---

(1) A similar theorem was stated by Sobolewski [9] in the special case where A(t) are a positive self-adjoint operator in a Hilbert space.

We again construct a sequence $\{A_n(t)\}$ of bounded operators that approximate $A(t)$. Here it is convenient to choose

(28) $$A_n(t) = A(t)J_n(t) , \quad J_n(t) = (1 + \frac{1}{n} A(t)^h)^{-m} ,$$

that is,

(28') $$A_n(t)^h = A(t)^h(1 + \frac{1}{n} A(t)^h)^{-1} .$$

It is easy to show, as in the case $h = 1$, that $A_n(t)$ satisfies the conditions i), ii), iii) with constants M independent of n. The evolution operators $U_n(t, s)$ for $A_n(t)$ can be constructed as before, and we want to show that $\text{s-lim}\, U_n(t, s) = U(t, s)$ exists.

To this end we deduce an expression, similar to (13), which is uniform in n. Again we use the identity (9), but the estimate (11) is no longer true for $m > 1$. At this point we use the following identity, essentially due to Sobolewski (see [9]).

(29) $$A_n(s) - A_n(r) = \sum_{p=1}^{m} A_n(s)^{1-ph}(A_n(s)^h A_n(r)^{-h} - 1) A_n(r)^{ph} .$$

If we put (29) into (10), the identity (9) becomes an integral equation involving the quantities

(30) $$X_{nq}(t, s) = U_n(t, s) A_n(s)^{1-qh} , \qquad q = 1, \dots , m$$

To get a closed system of integral equations for X_{nq}, we multiply (q) from right with $A_n(r)^{1-qh}$, $q = 1, \dots , m$. The result may be written symbolically

(31) $$X_{nq} = X_{nq}^{(o)} + \sum_{p=1}^{m} X_{np} * K_{npq} , \qquad q = 1, \dots , m ,$$

where

(32)
$$X_{nq}^{(o)}(t,s) = e^{-(t-s)A_n(s)} A_n(s)^{1-qh} = A_n(s)^{1-qh} e^{-(t-s)A_n(s)} ,$$
$$K_{npq}(t,s) = (A_n(t)^h A_n(s)^{-h} - 1) A_n(s)^{1+ph-qh} e^{-(t-s)A_n(s)}$$

The system (31) of integral equations can be solved by successive approximation :

(33)
$$X_{nq} = \sum_{k=o}^{\infty} X_{nq}^{(k)} , \quad X_{nq}^{(k)} = \sum_{p=1}^{m} X_{np}^{(k-1)} * K_{npq}$$

The question is whether there is a uniform (independent of n) majorizing series to (33) . Such a series exists since we have the uniform estimates

(34)
$$\left| X_{nq}^{(o)}(t,s) \right| \leq \frac{C}{(t-s)^{1-qh}} ,$$
$$\left| K_{npq}(t,s) \right| \leq \frac{C}{(t-s)^{1+ph-qh-\theta}} .$$

(34) follows from the fact that $-A_n(t)$ satisfies i), ii), iii). (note (27'')).

The strongest singularities on the right of (34) occurs for $q = 1$ and $p = m$. Then $1-qh = 1-h$ and $1+ph-qh-\theta = 2-h- \quad < 1$ by hypothesis. Thus these singularities are not too strong, and the series (33) converges uniformly in n , as we wished to show.

It follows, quite in the same way as in Theorem 2. 1. that s-lim $U_n(t-s) = U(t,s)$ exists and satisfies the conditions 1), 2) .

To prove the differentiability of $U(t,s)$ and the existence of $A(t)U(t,s) \in B$, we need another expression for $A_n(t)U_n(t,s)$ similar to (21). This can be obtained by deducing from (18) a system of integral equations satisfied by

(35)
$$Y_{nq}(t,s) = A_n(t)^{qh} U_n(t,s) , \qquad q = 1,\dots, m ,$$

and solving this system by successive approximation. For $q = m$, $Y_{nq}(t,s) = A_n(t)U_n(t,s)$ and this will give the desired differentiability of $U(t,s)$ as in the proof of Theorem 2.1. Here again we have a slight complication in constructing the first approximation $Y_{nq}^{(1)}$, similar to that encountered in (22). But the main feature of the proof is the same as before, and the details may be omitted.

6. Theorems 2.4. and 2.5. generalize the earlier theorems. Since, however, the fractional power $A(t)^h$ of a given operator is defined only in an abstract fashion and not given in a simple form, the question arises how one can verify the assumptions of these theorems.

There is a rather general case in which these assumptions are known to be satisfied. Let X and X' be two Hilbert spaces such that $X' \subset X$ (algebraically and topologically). Let $a(u,v)$ be a continuous sesquilinear form on X' such that $\operatorname{Re} a(u,u) \geq \delta |u|'^2$, $\delta > 0$ ($|u|'$ is the norm in X'), for all $u \in X'$. Then there exists a linear operator $-A \in (H0)$ such that $D_A \subset X'$ and $a(u,v) = (Au,v)$ for $u \in D_A$ and $v \in X'$ (cf. Lions' lectures). Now it can be shown that D_{A^α} is independent of A (depending only on X') for $0 \leq \alpha < 1/2$ (see [4]).

Consider next a family $a(t; u,v)$ of continuous sesquilinear forms on X' such that $\operatorname{Re} a(1; u,u) \geq \delta |u|'^2$. Let $A(t)$ be the operator associated with $a(t; u,v)$ in the way described above. Then $D_{A(t)^\alpha}$ is constant if $0 \leq \alpha < 1/2$. Suppose, moreover, that $a(t; u,v)$ is Hölder continuous with exponent ρ for each fixed $u, v \in X'$. Then it can be proved that $A(t)^\alpha A(0)^{-\alpha}$ is Hölder continuous with exponent ρ. Thus Theorem 2.4. is applicable to such a case with $m = 3$, $h = 1/3$ if $\rho > 2/3$ (see [4]).

Also it is very likely that the assumptions of Theorem 2.4. are satisfied when $A(t)$ is a family of strongly elliptic differential operator, if the region Ω , the boundary conditions and the coefficients are sufficiently regular in X as well as in t (see Remark 2. 6.).

7. Here we shall show that $Y_n^{(1)}(t,s)$ has the estimate (22). We note, once for all, that there is no question about the existence of $Y_n^{(1)}(t,s)$, for all functions such as $Y_n^o(t,s)$ and $H_n(t,s)$ are smooth at $t = s$ because $A_n(t) \in B$. The only question is to obtain a uniform estimate such as (22).

In this n° we shall omit the subscripts n and write $A(t)$, $Y(t)$, $H(t), \ldots$ in place of $A_n(t)$, $Y_n(t)$, $H_n(t), \ldots$ (so that $A(t) \in B$ in the following).

By definition we have

$$Y^{(1)} = H * Y^{(o)} = I_1 + I_2 + I_3 \ , \tag{36}$$

where

$$\begin{aligned} I_1(t,r) &= \int_r^t H(t,s)\left[A(s)e^{-(s-r)A(s)} - A(r)e^{-(s-r)A(r)}\right] ds \ , \\ I_2(t,r) &= \int_r^t H(t,r)\, A(r)e^{-(s-r)A(r)}\, ds \\ &= H(t,r)(1 - e^{-(t-r)A(r)}) \qquad \text{(see Lemma 1. 12)}, \\ I_3(t,r) &= \int_r^t \left[H(t,s) - H(t,r)\right] A(r)e^{-(s-r)A(r)} ds \ . \end{aligned} \tag{37}$$

We now need

Lemma 2. 7. Under the assumptions of Theorem 2. 1. we have

$$\left| A(t)e^{-\tau A(t)} - A(s)e^{-\tau A(s)} \right| \leq \frac{C}{|\tau|} (t-s)^{\varrho} \ ,$$

$$\left| A(t)(e^{-\tau A(t)} - e^{-\sigma A(t)}) \right| \leq \frac{C}{\beta} \frac{(\tau-\sigma)^{\beta}}{\sigma^{1+\beta}} \ , \qquad \begin{array}{l} 0 < \sigma < \tau \\ 0 < \beta \leq 1 \ . \end{array}$$

Proof. From the expression (3) of §1 for $-Ae^{-\tau A}$ we have

$$A(t)e^{-\tau A(t)} - A(s)e^{-\tau A(s)} = \frac{1}{2\pi i}\int e^{\lambda\tau}(\lambda + A(t))^{-1}(A(t)-A(s))(\lambda + A(s))^{-1}.\lambda\, d\lambda$$

Since $A(t) - A(s) = (A(t)A(s)^{-1} - 1)\, A(s)$ and $A(s)(\lambda + A(s))^{-1} =$ $= \left| 1 - \lambda\, (\lambda + A(s))^{-1} \right| \le 1 + M$, we have i) and by iii)

$$\left| A(t)e^{-\tau A(t)} - A(s)e^{-\tau A(s)} \right| \le C(t-s)^{\vartheta} \int \left| e^{\lambda\tau} \right| \frac{|\lambda|}{|\lambda|}\, d\lambda \le C\, \frac{(t-s)^{\vartheta}}{|\tau|} .$$

Again (in the proof of the second inequality we may write $A(t) = A$)

$$\left| A(e^{-\tau A} - e^{-\sigma A}) \right| = \left| \int_{\sigma}^{\tau} A^2 e^{-sA} ds \right| \le C \int_{\sigma}^{\tau} \frac{ds}{s^2}$$

(see (27'')). But (set $s = r^{1/\beta}$)

$$\int_{\sigma}^{\tau} \frac{ds}{s^2} = \frac{1}{\beta} \int_{\sigma^{\beta}}^{\tau^{\beta}} r^{-\frac{1}{\beta}-1}\, dr \le \frac{1}{\beta\sigma^{1+\beta}}(\tau^{\beta} - \sigma^{\beta}) \le \frac{(\tau - \sigma)^{\beta}}{\beta\sigma^{1+\beta}} ,$$

(q. e. d.)

By Lemma 2. 7. we can estimate I_1 :

$$\left| I_1(t,r) \right| \le C \int_r^t (t-s)^{-1+\vartheta}\, (s-r)^{-1+\vartheta}\, ds \le C(t-r)^{-1+2\vartheta} ,$$

It is easy to estimate I_2 :

$$\left| I_2(t,r) \right| \le C(t-r)^{-1+\vartheta}$$

I_3 is further divided into two integrals :

$$I_3(t,r) = I_3' + I_3'' \equiv \int_r^{\frac{r+t}{2}} + \int_{\frac{r+t}{2}}^{t}$$

where

$$I_3''(t,r) \le \int_{\frac{r+t}{2}}^{t} \left(|H(t,s)| + |H(t,r)| \right) \left| A(r)e^{-(s-r)A(r)} \right| ds$$

$$\le C \int_{\frac{r+t}{2}}^{t} \left[(t-s)^{-1+\vartheta} + (t-r)^{-1+\vartheta} \right] (s-r)^{-1} ds \le C(t-r)^{-1+\vartheta}$$

In estimating $I_3'(t,r)$, we use the identity :

$$H(t,s) - H(t,r) = A(t)\left[e^{-(t-s)A(t)} - e^{-(t-r)A(t)}\right](A(t)A(s)^{-1} - 1)$$
$$- A(t)\, e^{-(t-r)A(t)}\, A(t)A(s)^{-1}(A(s)A(r)^{-1} - 1)$$

Using the second inequality of Lemma 2. 7 with $\beta = \theta$, we have

$$|H(t,s)-H(t,r)| \leq \frac{C}{\theta}(t-s)^{-1-\theta}(s-r)^{\theta}(t-s)^{\theta} + C(t-r)^{-1}(s-r)^{\theta} \leq C(t-r)^{-1}(s-r)^{\theta}$$
$$\text{if } s \leq \frac{t+r}{2},$$

and hence

$$|I_3'(t,r)| \leq C \int_r^{\frac{r+t}{2}} (t-r)^{-1}(s-r)^{\theta}\,(s-r)^{-1} ds \leq C(t-r)^{-1+\theta} .$$

Collecting the above results, we obtain (22).

§ 3. THE CASE IN WHICH $-A(t)$ ARE GENERATORS OF ANALYTIC SEMI-GROUPS WITH VARIABLE DOMAIN.

1. In this section we prove a theorem, due to Tanabe[1)], in which no assumption is made on the constancy of $D_{A(t)}$ or of $D_{A(t)^h}$.

Theorem 3. 1. Assume that

i) $-A(t) \in (H0)$ uniformly in the sense of Theorem 2. 1., i).

ii) $A(t)^{-1}$ is differentiable, with $\frac{d}{dt} A(t)^{-1}$ Hölder continuous (in norm). This implies that the derivative exists in norm. Also this implies that

$$\frac{d}{dt}(\lambda + A(t))^{-1} = \frac{A(t)}{\lambda + A(t)} \left(\frac{d}{dt} A(t)^{-1}\right) \frac{A(t)}{\lambda + A(t)} = \tag{1}$$

$$= \left(1 - \frac{\lambda}{\lambda + A(t)}\right)\left(\frac{d}{dt} A(t)^{-1}\right)\left(1 - \frac{\lambda}{\lambda + A(t)}\right)$$

exists and is analytic in λ.

iii)
$$\left| \frac{d}{d\lambda}(\lambda + A(t))^{-1} \right| \leq \frac{M}{|\lambda|^{1-\rho}}$$

for some constants M, ρ, such that $0 \leq \rho < 1$.

Then there exists a unique evolution operator $U(t, s)$ with the properties stated in Theorem 2. 1. (including the case in which $A(t)^{-1}$ is analytic).

Remark 3. 2. If one is satisfied with construncting $U(t, s)$ which is only weakly differentiable, one may omit the assumption of Hölder continuity of $\frac{d}{dt} A(t)^{-1}$ (such $U(t, s)$ is still unique).

---

(1) See [13] and [6]. The proof given below is slightly different from the original.

Proof of Theorem 3. 1. Again we use the approximating sequence $A_n(t)$ used in the proof of Theorem 2. 1. and construct the approximating evolution operators $U_n(t, s)$.

To obtain a uniform expression for $U_n(t, s)$, however, we make slight modifications in the method used in Theorem 2. 1.

Instead of (9) of § 2, we deduce a different integral equation for U_n, starting from the identity

(2) $$\frac{\partial}{\partial s} e^{-(t-s)A_n(s)} U_n(s,r) = K_n(t-s;s)\, U_n(s,r),$$

where

(3) $$K_n(\tau;s) = \frac{\partial}{\partial s} e^{-\tau A_n(s)} .$$

Note that two other derivatives in (2) cancel each other since $\frac{\partial}{\partial s} U_n(s,r) = -A_n(s)U_n(s,r)$ and $\frac{\partial}{\partial \tau} e^{-\tau A_n(s)} = -e^{-\tau A_n(s)} A_n(s)$.

On integration with respect to s on (r, t), (2) gives the integral equation

(4) $$U_n = U_n^{(o)} + \hat{K}_n * U_n$$

where

(5) $$\begin{cases} U_n^{(o)}(t,s) = e^{-(t-s)A_n(s)} \\ \hat{K}_n(t,s) = K_n(t-s;\, s) \end{cases}$$

($U_n^{(o)}$ and K_n differ from the operators $U_n^{(o)}$ and K_n used in the proof of Theorem 2. 1.).

Now it can be proved that

(6) $$\left| \hat{K}_n(t,s) \right| \leq \frac{C}{(t-s)^{\rho}}$$

where C is independent of n (for the proof see n. 2). Since $\rho < 1$, (4) can be solved by a successive approximation used in the proof of Theorem 2. 1., yielding a series expression of $U_n(t,s)$ converging uniformly in n. It follows by the same argument that $\text{s-lim}_{n\to\infty} U_n(t,s) = U(t,s)$ exists and has the properties 1), 2) of Theorem 2.1.

To prove 3), we use another integral equation obtained from

$$\frac{\partial}{\partial s} U_n(t,s)e^{-(s-r)A_n(s)} = U_n(t,s)K_n(s-r;s) .$$

After integration and multiplication from left by $A_n(t)$, this gives the following integral equation for $Y_n(t,s) = A_n(t)\,U_n(t,s)$:

$$Y_n = Y_n^{(o)} + Y_n * \hat{H}_n \quad , \tag{7}$$

where

$$\begin{cases} Y_n^{(o)}(t,s) = A_n(t)\, e^{-(t-s)A_n(t)} , \\ \hat{H}_n(t,s) = K_n(t-s;t) . \end{cases} \tag{8}$$

Again we have the uniform estimate (see n. 2)

$$\left| \hat{H}_n(t,s) \right| \leq \frac{C}{(t-s)^{\rho}} . \tag{9}$$

For $Y_n^{(o)}$, however, we have only the estimate $\left| Y_n^{(o)}(t,s) \right| \leq C(t-s)^{-1}$ of which the right member is not integrable. Thus the estimate of $Y_n^{(1)}(t,s)$ is not quite simple. If one uses the Hölder continuity of $\frac{d}{dt} A(t)^{-1}$, however, we can carry out an estimation of $Y_n^{(1)}$, obtaining

$$\left| Y_n^{(1)}(t,s) \right| \leq \frac{C}{(t-s)^{\rho}} + \frac{C}{(t-s)^{1-\theta}} \tag{10}$$

where C is independent of n and θ is the exponent of the Hölder continuity of $\frac{d}{dt} A_n(t)^{-1} = \frac{d}{dt} A(t)^{-1}$ (see (5) of § 2).

Since the right member of (10) is integrable, there is no further difficulty in proceeding with higher approximations $Y_n^{(k)}$, $k = 2, 3, \dots$. The remaining arguments are exactly the same as in the proof of Theorem 2.1. and may be omitted , q. e. d.

2. Proof of (6) and (9).

It follows from (6) of § 2 that

$$\frac{\partial}{\partial t}(\lambda + A_n(t))^{-1} = \frac{n^2}{(n+\lambda)^2} \frac{\partial}{\partial t}\left(\frac{n\lambda}{n+\lambda} + A(t)\right)^{-1}.$$

Hence by iii)

$$\left| \frac{\partial}{\partial t}(\lambda + A_n(t))^{-1} \right| \leq \frac{n^2}{|n+\lambda|^2} \cdot M\left(\frac{|n+\lambda|}{|n\lambda|}\right)^{1-\rho} =$$

$$= M\left|\frac{n}{n+\lambda}\right|^{1+\rho} \frac{1}{|\lambda|^{1-\rho}} \leq \frac{C}{|\lambda|^{1-\rho}}$$

where C is independent of n.

The last inequality has the same form as iii) for $A(t)$. Therefore, we may prove, instead of (6) and (9), the same estimates for $\hat{K}$ and $\hat{H}$, omitting the subscript n.

From (1) of § 1 we have

$$K(\tau; s) \equiv \frac{\partial}{\partial s} e^{-\tau A(s)} = \frac{1}{2\pi i} \int_C e^{\lambda\tau} \frac{\partial}{\partial s}(\lambda + A(s))^{-1} d\lambda .$$

Hence

$$|K(\tau; s)| \leq \frac{M}{2\pi} \int_C |e^{\lambda\tau}| \frac{|d\lambda|}{|\lambda|^{1-\rho}} \leq \frac{C}{|\tau|^{\rho}}$$

as in the proof of (3') of § 1. Since $\hat{K}(t, s) = K(t-s; s)$ and $\hat{H}(t, s) = K(t-s; t)$, (6) and (9) follow immediately.

§4. THE CASE IN WHICH $-A(t)$ ARE CONTRACTION SEMI-GROUPS.

1. So far we have been considering only the cases $-A(t) \in (H0)$. If we assume only that $-A(t) \in (B0)$, there arise several difficulties. In this case we can still construct the approximating sequence $\{A_n(t)\}$ and the associated evolution operators $U_n(t,s)$, but the uniform expression such as (13) of § 2 is difficult to obtain, for it is essentially tied up with the property that $-A(t) \in (H0)$.

It is easy to prove that

$$U_n(t,r) - U_m(t,r) = \int_r^t U_m(t,s)(A_m(s)-A_n(s))U_n(s,r)ds,$$

but it appears to be very difficult to prove the strong convergence of U_n from this identity, although this is the very identity used in the construction of $U(t,r) = e^{-(t-r)A}$ when $A(t) = A$ is independent of t (see Phillips) In that case $A_m - A_n$ commutes with $U_n(s,r) = e^{-(s-r)A_n}$, a fact essential in that construction.

Here we will state a theorem in which, among others, we assume that $-A(t) \in (C0)$ (generator of a contraction semi-group). No satisfactory theorem is known in the general case of (B0).

<u>Theorem 4. 1.</u> Assume that

i) $-A(t) \in (C0)$, that is, $|(\lambda + A(t))^{-1}| \leq \frac{1}{\lambda}$, $\lambda > 0$. Also we assume (for convenience) that $A(t)^{-1} \in B$, $|A(t)^{-1}| \leq M$.

ii) $D_{A(t)} = D = \text{const}$ (again this implies that $A(t)A(s)^{-1} \in B$).

iii) $B(t) = A(t)A(o)^{-1} \in B$ is strongly continuously differentiable.

(This implies that $B(t)$ is continuous in norm, so that, since $B(t)^{-1} = A(o)A(t)^{-1} \in B$ by ii),

$$|B(t)| \leq M, \qquad |B(t)^{-1}| \leq M).$$

Then there exists a unique evolution operator $U(t,s)$ such that

1) $U(t,s) \in B$, strongly continuous, for $0 \leq s \leq t \leq T$.

2) $U(t,s)U(s,r) = U(t,r)$, $U(t,t) = 1$.

3) $A(t)U(t,s)A(s)^{-1} = W(t,s) \in B$, strongly continuous in s,t.

4) $U(t,s)A(s)^{-1}$ is strongly continuously differentiable in t for $t \geq s$, with

$$\frac{\partial}{\partial t} U(t,s)A(s)^{-1} = A(t)U(t,s)A(s)^{-1} \in B .$$

Remark 4.2. 3) implies that $U(t,s)D \subset D$, and 4) implies that

$$\frac{\partial}{\partial t} U(t,s)u = A(t)U(t,s)u , \quad u \in D$$

Theorem 4.3. Under the assumptions of Theorem 4.1., the results of Theorem 1.11. are true (with (E_o) and (S_o) replaced by (E) and (S), respectively).

2. It is not possible to give here a complete proof of these theorems. We shall give a sketch of the proof of Theorem 4.1., referring for details to [1]. The proof of Theorem 4.2. is similar to that of Theorem 1.11.

The idea of the proof is to use the approximation of $A(t)$ by step functions. This is equivalent to approximate $U(T,0)$ by a product

$$(1) \qquad U_\Delta(T,0) = e^{-(t_n - t_{n-1})A_n} \cdots e^{-(t_1 - t_o)A_1}$$

where $A_j = A(t_j)$ and Δ denotes the partition $0 = t_o < t_1 < \ldots < t_n = T$ of the interval $[0, T]$. For simplicity we introduce the following notations

$$(2) \qquad \begin{cases} X_j = e^{-(t_j - t_{j-1})A_j} , & B_j = B(t_j) \\ U_{jk} = X_j X_{j-1} \cdots X_k , & W_j = A_j U_{j1} A_o^{-1} \end{cases}$$

Since $-A_j \in (C0)$, we have $\| X_j \| \leq 1$ and $\| U_{jk} \| \leq 1$. It can also be shown that W_j are uniformly bounded. To see this we use the recurrence formula

(3) $$W_j = U_{j1} + \sum_{k=1}^{j} U_{jk} (B_k - B_{k-1}) B_{k-1}^{-1} W_{k-1} ,$$

which can be verified in a straightforward way if one notes the commutativity of A_k and X_k : $A_k X_k \supset X_k A_k$.

(3) is a recurrence equation of Volterra type, and can be solved by successive approximation (actually in a finite number of steps). The solution may be written

(4) $$W_j = \sum_{p=o}^{\infty} W_j^{(p)} , \qquad W_j^{(o)} = U_{j1}$$
$$W_j^{(p+1)} = \sum_{k=1}^{j} U_{jk} (B_k - B_{k-1}) B_{k-1}^{-1} W_{k-1}^{(p)} .$$

Since $|B_{k-1}^{-1}| \leq M$ by iii) and

(5) $$\sum_{k=1}^{n} |B_k - B_{k-1}| \leq \int_o^T |dB(t)| \leq \int_o^T |B'(t)| dt \equiv N ,$$

it is easy to see that the series (4) for W_j converges uniformly (with respect to the partition Δ , if one supposes that the sequence $\{W_j\}$ has been extended to a step function defined for all t). In particulat W_j are uniformly bounded :

(6) $$|W_j| \leq C = \text{const.}$$

3. We now show that $U_\Delta (T, 0)$ has a strong limit for

$$|\Delta| \equiv \max_j |t_j - t_{j-1}| \longrightarrow 0 .$$

To this end we consider a partition Δ' of $[0, T]$ and a subpartition Δ of Δ'. Let Δ be $0 = t_o < t_1 < \dots < t_n = T$. The partition points of Δ' form a subset of $\{t_o, t_1, \dots, t_n\}$. By using the semi-group property of $\{e^{-\tau A_j}\}$, however we can write

$$U_\Delta = U_\Delta(0,T) = X_n, \ldots, X_1, \qquad X_j = e^{-(t_j - t_{j-1})A_j},$$

$$U_{\Delta'} = U_{\Delta'}(0,T) = X'_n, \ldots, X'_1, \qquad X'_j = e^{-(t_j - t_{j-1})A'_j},$$

with the same number n of factors.

Here $A_j = A(t_j)$, $A'_j = A_j(t'_j)$, where t'_j is equal to some t_k such that $| t'_j - t_k | \leq | \Delta' |$

We have then

$$(U_{\Delta'} - U_\Delta)A_o^{-1} = \sum_{j=1}^{n} U'_{n,j+1}(X'_j - X_j)U_{j-1,1}A_o^{-1}$$

$$= \sum U'_{n,j+1}(X'_j - X_j)A_{j-1}^{-1} W_{j-1}$$

where $U_{o1} = 1$, $U'_{n,n+1} = 1$ by convention. Hence

$$(7) \qquad |(U_{\Delta'} - U_\Delta)A_o^{-1}| \leq C \sum |(X'_j - X_j)A_{j-1}^{-1}|$$

because $|U'_{n,j+1}| \leq 1$ and $|W_{j-1}| \leq C$.

It follows from Lemma 4. 4. (to be proved below) that

$$|(X'_j - X_j)A_{j-1}^{-1}| \leq |(X'_j - X_j)A_j^{-1}| \; |A_j A_{j-1}^{-1}|$$

$$\leq M(t_j - t_{j-1}) |(A'_j - A_j)A_j^{-1}|$$

$$\leq M(t_j - t_{j-1}) |(B'_j - B_j)B_j^{-1}|$$

$$\leq M^2(t_j - t_{j-1}) | B'_j - B_j | .$$

Summing this over subintervals in Δ of a fixed subinterval I'_ν of Δ' gives a quantity $\leq M^2 \left(\int_{I'} dB(t) \right) \sum (t_j - t_{j-1}) \leq M^2 |\Delta'| \int_{I'_\nu} dB(t)$.

Summing the results over all I'_ν we get a quantity $\leq M^2 |\Delta'| \int_0^T dB(t) =$
$= M^2 N |\Delta'|$.

Hence by (7)

(8) $$\left| (U_{\Delta'} - U_{\Delta}) A_0^{-1} \right| \leq C |\Delta'|$$

Now let Δ' and Δ'' be arbitrary partitions of $[0, T]$. If Δ is their common subpartition, we have (8) and a similar inequality with Δ' replaced by Δ''. Hence

(9) $$\left| (U_{\Delta'} - U_{\Delta''}) A_0^{-1} \right| \leq C(|\Delta'| + |\Delta''|).$$

This shows clearly that $\lim U_\Delta A_0^{-1}$ exists <u>in norm</u> when $|\Delta| \to 0$. Since U_Δ is uniformly bounded ($|U_\Delta| \leq 1$) and D_{A_0} is dense, it follows that $U = U(T,0) = s - \lim_{|\Delta|\to 0} U_\Delta (T,0)$ exists.

In the same way we can construct $U(t,s) = s - \lim_{|\Delta|\to 0} U_\Delta (t,s)$ by considering partitions Δ of $[s,t]$.

The relation $U(t,s)U(s,r) = U(t,r)$ then follows immediately from $U_{\Delta'}(t,s)U_{\Delta''}(s,r) = U_\Delta (t,r)$ by going to the limit, when Δ', Δ'' are partitions of $[s,t]$, $[r,s]$, respectively, and Δ is the partition of $[r,t]$ obtained by joining Δ' and Δ''.

Incidentally we note

(9a) $$\left| (U_\Delta - U) A_0^{-1} \right| \leq M^2 |\Delta| \int_0^T |B'(t)| dt .$$

which follows from (9) by letting $|\Delta''| \to 0$, $\Delta' = \Delta$

To prove the continuity of $U(t,s)$, it is convenient to introduce the notation

$$U(t,s;\Delta) = e^{-(t-t_{j-1})A_j} e^{-(t_{j-1}-t_{j-2})A_{j-1}} \ldots\ldots e^{-(t_k-s)A_k}$$

$$\text{if} \quad t_{j-1} \leq t \leq t_j, \quad t_{k-1} \leq s \leq t_k,$$

where Δ is a given partition of $[0, T]$. Then it is easy to show that

(10) $\quad \left\| \left[U(t,s;\Delta) - U(t,s) \right] A_0^{-1} \right\| \leq C \, |\Delta|$

so that

(11) $\quad \left\| \left[U(t,s;\Delta) - U(t,s) \right] A_0^{-1} \right\| \longrightarrow 0, \qquad |\Delta| \longrightarrow 0.$

Thus $U(t,s)A_0^{-1}$ is the uniform (in s, t) limit in norm of $U(t,s;\Delta)A_0^{-1}$ for $|\Delta| \to 0$, so that $U(t,s)$ is the strong limit of $U(t,s;\Delta)$ (uniformly in s, t). Since $U(t,s;\Delta)$ is strongly continuous in s, t, the same is true with $U(t,s)$, with $U(t,t) = 1$ (since $U(t,t;\Delta) = 1$).

4. To prove 3), we return to the expression (4) for W_j. After extending $\{W_j\}$ and $\{W_j^{(p)}\}$ as step functions for all values of t (as mentioned after (5)), we can prove by riduction that each term $W_j^{(p)}$ tends strongly to a function $W^{(p)}(t)$ for $|\Delta| \longrightarrow 0$. In view of uniform convergence of (4) with respect to Δ, it follows that $W_j \longrightarrow W(t) = \sum_{p=0}^{\infty} W^{(p)}(t)$. In particular $W_\Delta = W_\Delta(T,0) = W_n = A(T)U_\Delta(T,0)A_0^{-1} \longrightarrow W = W(T,0)$. Hence $U(T,0)A_0^{-1} = \lim_{\Delta \to 0} U_\Delta(T,0)A_0^{-1} = \lim A(T)^{-1}W_\Delta = A(T)^{-1}W(T,0)$ so that

$$A(T)U(T,0)A_0^{-1} = W(T,0) \in B$$

This is a special case of 3). The general case and the continuity of $W(t,s)$ can be dealt with in the same way.

To prove the differentiability 4), we note that

$$U(t+\varepsilon,s) - U(t,s) = (U(t+\varepsilon,t) - 1)U(t,s),$$

$$(U(t+\varepsilon,s) - U(t,s))A(s)^{-1} = (U(t+\varepsilon,t)-1)A(t)^{-1}W(t,s),$$

where $W(t,s) = A(t)U(t,s)A(s)^{-1} \in B$. To prove 4) it suffices to prove that

$$\frac{1}{\varepsilon}(U(t+\varepsilon,s) - 1)A(t)^{-1} \longrightarrow -1, \qquad \varepsilon \downarrow 0.$$

Now we have (apply (9a) to the trivial partition $t = t_0 < t_1 = t+\varepsilon$ of $[t, t+\varepsilon]$)

$$\left| (U(t+\varepsilon, t) - e^{-\varepsilon A(t)})A(t)^{-1} \right| \leq C\,\varepsilon \int_t^{t+\varepsilon} \left| B'(s) \right| ds .$$

Hence

$$\frac{1}{\varepsilon} \left| (U(t+\varepsilon, t) - e^{-\varepsilon A(t)})A(t)^{-1} \right| \to 0 , \qquad \varepsilon \downarrow 0 .$$

But

$$\left| \frac{1}{\varepsilon} (e^{-\varepsilon A(t)} - 1)A(t)^{-1} + 1 \right| \to 0$$

by Theorem 1.2.. Hence

$$\left| \frac{1}{\varepsilon} (U(t+\varepsilon, t) - 1) A(t)^{-1} + 1 \right| \to 0 , \qquad \varepsilon \downarrow 0$$

as we wished to show.

Actually this proves only that

$$\left(\frac{\partial}{\partial t}\right)^{+} U(t, s)A(s)^{-1} = W(t, s)$$

when $\left(\frac{\partial}{\partial t}\right)^{+}$ denotes the right derivative. But since $W(t, s)$ is known to be strongly continuous, it follows that $\frac{\partial}{\partial t} U(t, s)A(s)^{-1}$ exists and equals $W(t, s)$.

5. Remarks.

a) It is not easy to extend Theorem 4.1. to the case $-A(t) \in (B0)$, for $|X_j| \leq 1$ is used essentially. Also it is not easy to weaken the assumption that $D_{A(t)}$ is constant. However, these generalizations are possible under certain circumstances.

Suppose that there is an operator $Q(t) \in B$, depending on t smoothly together with $Q(t)^{-1} \in B$, such that $-Q(t)^{-1}A(t)Q(t) \in (C0)$. Suppose furthermore, that there is another operator $R(t) \in B$, again depending smoothly on t together with $R(t)^{-1} \in B$, such that $R(t)^{-1}A(t)R(t)$ has a constant domain and a continuity property similar to iii) required of $A(t)$ there. Then Theorems 4. 1. and 4. 2. are seen to hold (see [2]).

b) In the beginning of this section, we stated that it is difficult to prove the existence of $\lim_{n\to\infty} U_n(t,s)$. Once the existence of $U(t,s)$ has been established as in Theorem 2. 1., however, it is not difficult to prove that $U_n(t,s) \to U(t,s)$ strongly.

c) The method used in the proof of Theorem 4. 1. is analogous to the <u>difference approximation</u> to the differential equation, but not exactly the same.

The true difference method would be to use the scheme

$$\frac{u(t_j) - u(t_{j-1})}{t_j - t_{j-1}} + A(t_j)u(t_j) = 0 \tag{12}$$

(we set $f = 0$ for simplicity), choosing the backward difference. This gives

$$u(t_j) = (1 + (t_j - t_{j-1})A_j)^{-1} u(t_{j-1}), \qquad A_j = A(t_j)$$

and so

$$u(t_j) = (1 + (t_j - t_{j-1})A_j)^{-1} \ldots\ldots\ (1 + (t_1 - t_o)A_1)^{-1} u_o \ .$$

It is expected, then, that

$$\lim_{|\Delta|\to 0} (1 + (t_n - t_{n-1})A_n)^{-1} \ldots\ldots\ (1 + (t_1 - t_o)A_1)^{-1} = U(T,0) \ . \tag{13}$$

This means that we should be able to replace $X_j = e^{-(t_j - t_{j-1})A_j}$ used in the above proof by $(1 + (t_j - t_{j-1})A_j)^{-1}$, which is an operator of more elementary character than X_j. But the direct proof of the existence of the limit (13) is not known, although it can be proved after one has proved the existence of $U(t,s)$.

BIBLIOGRAPHY

[1] T. KATO, J. Math. Soc. Japan 5 (1953), 208-234.

[2] T. KATO, Comm. Pure Appl. Math. 9 (1956), 479-486.

[3] T. KATO, Proc. Japan Acad. 36 (1960), 94-96.

[4] T. KATO, J. Math. Soc. Japàn 13 (1961), 246-274.

[5] T. KATO, Nagoya J. Math. 19 (1961), 93-125.

[6] T. KATO and H. TANABE, Osaka Math. J. 14 (1962), 107-133.

[7] H. KOMATSU, J. Fac. Sci. Univ. Tokyo, Sec. I, Vol. 9, Part 1 (1961), I-II.

[8] R. S. PHILLIPS, Trans. Amer. Math. Soc. 74 (1954), 199-221.

[9] P. E. SOBOLEWSKI, Doklady Akad. Nauk U. S. S. R. 123 (1958), 984-987.

[10] H. TANABE, Osaka Math. J. 11 (1959), 121-145.

[11] H. TANABE, Osaka Math. J. 12 (1960), 145-166.

[12] H. TANABE, Osaka Math. J. 12 (1960), 363-376.

[13] H. TANABE, Proc. Japan Acad. 37 (1961), 610-613.

[14] H. TANABE, Technical Report, NSF, G-22982, Dept. Math. , Univ. California, 1963.

[15] K. YOSIDA, Proc. Japan Acad. 34 (1958), 337-340.

[16] K. YOSIDA, Proc. Japan Acad. 36 (1960), 86-89.

[17] K. YOSIDA, Berkeley Sympsium, 1960.

CENTRO INTERNAZIONALE MATEMATICO ESTIVO

(C.I.M.E.)

J.L.LIONS

ÉQUATIONS DIFFÉRENTIELLES OPÉRATIONELLES DANS LES ESPACES DE HILBERT

ROMA - Istituto Matematico dell'Università

ÉQUATIONS DIFFÉRENTIELLES OPÉRATIONELLES DANS LES ESPACES DE HILBERT

par

J. L. LIONS

Soit H un espace de Hilbert, t un paramètre réel, dans un intervalle (o, T) ; et A(t) une famille d'opérateurs non _bornés_ dans H .

On appelle _équation différentielle opérationelle_ dans H une équation de la forme :

$$A(t)\,u(t) + \frac{du(t)}{dt} \quad (\text{ou} + \frac{d^2u}{dt^2}) = f(t), \qquad f \text{ donné} \tag{1}$$

avec

$$u(o) \text{ donné} \quad (\text{ou } u(o) \text{ et } \frac{du(o)}{dt} \text{ donnés})$$

et u(t) étant dans le domaine de A(t) pour que (1) ait un sens.

Le chapitre I donne les outils permettant de définir des opérateurs non bornés commodes pour notre objet et qui couvrent un très grand nombre d'opérateurs différentielles avec des conditions aux limites variées (pour des _exemples,_ nous renvoyons à notre ouvrage: Équations différentielles opérationelles et problèmes aux limites, Springer, Collection Jaune, t. 111(1961)).

Le chapitre II donne quelques résultats pour les équations différentielles du 1er ordre (i.e. contenant seulement $\frac{du}{dt}$; A(t) peut être un opérateur - ou un système - différentiel d'ordre quelconque) et contient une _introduction_ très brève aux problèmes non homogènes (les équations d'_évolution_ non homogènes sont étudiées par E. Magenes et l'A. dans une note aux C.R. Acad. Sc., Paris, t. 251, 2118- 2120, (1960), et dans un travail aux Rend. dei Lincei, 1963).

Le chapitre III contient, après quelques indications générales sur les problèmes non linéaires, un exposé, dans un cas particulier assez significatif, d'un travail de I.M. Visik, Mat. Sbornik, t.59(101), p.289-325 (1962).

Le chapitre IV revient aux problèmes linéaires, contenant cette fois des dérivées du 2ème ordre en t.

Nous avons insisté sur la régularité en t de la solution, ainsi que sur la dépendance en les coefficients.

Les chapitres I, II, IV ont évidemment des points communs avec notre ouvrage, loc.cit. Mais nous avons essayé de présenter ici soit des résultats nouveaux par rapport à ceux de notre livre, soit avec des méthodes nouvelles des résultats deja établis dans notre livre. (Des indications bibliographiques et des remarques sur les méthodes ou les problèmes non résolus sont données dans les commentaires placés à la fin de chaque chapitre.)

TABLE DES MATIÈRES

Chapitre III. Introduction à certains problèmes non linéaires.

Chapitre IV . Équations linéaires du 2ème ordre.

Chapitre I

COUPLES D'ESPACES HILBERTIENS

1. Hypothèses. Exemples

1.1 - On désigne par V et H deux espaces de Hilbert.

Si $f, g \in H$, (f, g) est leur produit scalaire dans H , $|f| = (f, f)^{1/2}$.

Si $u, v \in V$, $((u, v))$ est leur produit scalaire dans V, $\|u\| = ((u, u))^{1/2}$.

On suppose que $V \subset H$, l'injection de V dans H étant continue. Donc il existe une costante c telle que

$$(1.1) \qquad |u| \leq c \|u\| \qquad \text{pour tout} \quad u \in V.$$

On suppose enfin que V est dense dans H .

1.2 - Exemples.

Si Ω est un ouvert de R^n , on prend $H = L^2(\Omega)$ (classes de fonctions de carré sommable sur Ω pour la mesure de Lebesgue $dx = dx_1 \ldots\ldots dx_n$).

On désigne par $H^m(\Omega)$ l'espace de Sobolev d'ordre m : "$u \in H^m(\Omega)$" équivalant à

$$\text{"} D^p u \in L^2(\Omega) \quad \text{pour tout } p \text{ avec } |p| \leq m \text{"}$$

Ici

$$D^p u = \frac{\partial^{p_1 + p_2 + \cdots + p_n}}{\partial x_1^{p_1} \partial x_2^{p_2} \cdots \partial x_n^{p_n}} u , \qquad |p| = p_1 + \ldots + p_n \quad ,$$

et $D^p u$ est calculée au sens des distributions sur Ω .

Si l'on pose :

(1.2) $$|u|_m = \Big(\sum_{|p| \leq m} \int_\Omega |D^p u|^2 \, dx \Big)^{1/2} ,$$

on définit sur $H^m(\Omega)$ <u>une structure d'espace de Hilbert.</u>

On désigne par $\mathcal{D}(\Omega)$ l'espace des fonctions indéfiniment différentiables sur Ω et à support compact dans Ω (muni, lorsqu'il y a lieu, de la topologie de Schwartz).

<u>On désigne par</u> $H_0^m(\Omega)$ <u>l'adhérence de</u> $\mathcal{D}(\Omega)$ <u>dans</u> $H^m(\Omega)$.

Si Ω est un ouvert de frontière Γ suffisamment régulière, $H_0^m(\Omega)$ consiste en le sous espace de $H^m(\Omega)$ formé des u telles que:

(1.3) $$D^p u = 0 \quad \text{sur } \Gamma \quad \text{pour } |p| \leq m-1.$$

(Nous n'entendons pas étudier ici à quel sens (1.3) a lieu).

Si $\complement\Omega$ est "tres petit", alors $H^m(\Omega) = H_0^m(\Omega)$. Par ex. on pourra vérifier que $H^m(R^n) = H_0^m(R^n)$ (régularisation et troncature).

On peut alors prendre:

(1.4) $$V = H^m(\Omega) \qquad (\text{avec } \|u\| = |u|_m),$$

ou bien

(1.5) $$V = H_0^m(\Omega)$$

où, plus généralement, un sous espace fermé quelconque de $H^m(\Omega)$, contenant $H_0^m(\Omega)$. On vérifie sans peine que dans tous ces cas, les hypothèses du point 1.1 on lieu.

<u>Autre exemple.</u>

Voici un autre exemple (utile) et où V <u>n'est pas</u> un espace de Sobolev.

On designe par V l'espace des (classes de) fonctions u telle que

(1.6) $$u \in L^2(\Omega), \qquad \Delta u \in L^2(\Omega),$$

où $$\Delta = \frac{\partial^2}{\partial x_1^2} + \ldots\ldots + \frac{\partial^2}{\partial x_n^2} \, .$$

Muni de la norme

$$\left\{ \int_\Omega (|u|^2 + |\Delta u|^2) dx \right\}^{1/2}$$

c'est un espace de Hilbert, et les hypothèses du point 1.1 ont lieu.

On verifiera facilement, en prenant par ex. pour Ω un disque dans le plan et $\Delta u = 0$, que (1.6) n'entraine pas en général que les dérivées premières $\frac{\partial u}{\partial x_i}$ soient dans $L^2(\Omega)$; V n'est pas un espace de Sobolev.

2. L'espace V'. Exemples.

2.1. On désigne par V' l'anti-dual de V, i.e. l'espace des formes anti-linéaires continues sur V.

Si $f \in H$, $v \to (f, v)$ est une forme anti-linéaire continue sur V, donc definit un élément L_f de V' ; l'applications $f \to L_f$ est biunivoque de $H \to V'$ (car V est dense dans H). On identifie alors L_f à f. Donc:

(2.1) $$V \subset H \subset V'.$$

Si $f \in V'$, $v \in V$, (f, v) designera leur produit scalaire - dans l'anti-dualité. L'espace V' est un Hilbert pour la norme

$$\| f \|_{V'} = \sup \frac{|(f, v)|}{\| v \|}, \quad v \in V.$$

Naturellement, si $u \in V$, $v \to ((u,v))$ est continue sur V, donc definit un élément Λu de V' :

(2.2) $$((u,v)) = (\Lambda u, v).$$

On définit ainsi Λ avec:

(2.3) $$\Lambda \in \mathcal{L}(V;V')$$

(où, de façon générale, $\mathcal{L}(X;Y)$ désigne l'espace des applications linéaires continues de X dans Y); Λ <u>est un isomorphisme de</u> V <u>sur</u> V' (isomorphisme canonique).

Notons que

(2.4) $$\|f\|_{V'} = \|\Lambda^{-1} f\|$$

Si $u, v \in V$, on à évidemment (Cauchy-Schwarz)

(2.5) $$|(u,v)| \leq |u| \cdot |v| \quad .$$

On a aussi

(2.6) $$|(u,v)| \leq \|u\| \cdot \|v\|_{V'} \qquad (\text{ou } \|u\|_{V'} \cdot \|v\|).$$

2.2-<u>Exemples</u>.

Si nous prenons les exemples du point 1.2, avec $V = H_o^m(\Omega)$, alors V' est <u>un espace de distributions sur</u> Ω (grâce au fait que $\mathcal{D}(\Omega)$ est - par définition - dense dans $H_o^m(\Omega)$, on peut identifier V' à un sous espace de $\mathcal{D}'(\Omega)$ = distributions sur Ω). On posera :

(2.7) $$(H_o^m(\Omega))' = H^{-m}(\Omega);$$

$H^{-m}(\Omega)$ est composé des distributions T sur Ω de la forme

(2.8) $$T = \sum_{|p| \leq m} D^p f_p \quad , \qquad f_p \in L^2(\Omega),$$

la décomposition (2.8) n'étant pas unique.

On vérifiera que dans ce cas l'isomorphisme Λ donne:

$$u \longrightarrow \sum_{|p|\leq m} (-1)^{|p|} D^{2p} u$$

est un isomorphisme de $H_0^m(\Omega)$. sur $H^{-m}(\Omega)$. C'est une formulation (faible) du problème de Dirichlet pour l'opérateur $\sum_{|p|\leq m} (-1)^{|p|} D^{2p}$.

Autre exemple.

Si l'on prend $V = H^m(\Omega)$, alors V' n'est pas un espace de distributions sur Ω .

3. Une propriété du triplet $\{V, H, V'\}$.

3.1. Le triplet $\{V, H, V'\}$ possède la propriété suivante:

Théorème 3.1. Si π est dans l'espace $\mathcal{L}(V';V')$ (de norme $\|\pi\|'$) et a la propriété d'appliquer continûment V dans lui même (i.e. $\pi \in \mathcal{L}(V;V)$, de norme $\|\pi\|$), alors $\pi \in \mathcal{L}(H;H)$ et sa norme $|\pi|$ dans cet espace vérifie:

$$|\pi| \leq \max (\|\pi\|, \|\pi\|') .$$

Nous ne démontrerons pas ici ce théorème (cf. commentaires) mais nous en démontrerons un cas particulier, dû à P.D. Lax.

Enonçons d'abord ce cas particulier:

Théorème 3.2. Soit $\pi \in \mathcal{L}(V;V)$, de norme $\|\pi\|$, tel que

(3.1.) $$(\pi u, v) = (u, \pi v) \qquad \text{pour tout } u, v \in V.$$

Alors $\pi \in \mathcal{L}(H;H)$ et $|\pi| \leq \|\pi\|$.

Vérifions d'abord que ce théorème est un cas particulier du Théorème 3.1..

En effet, si $\Pi \in \mathcal{L}(V;V)$, alors son adjoint $\Pi^* \in \mathcal{L}(V';V')$ et (3.1) signifie que Π^* prolonge Π. Donc Π a les propriétés du Théorème 3.1, avec $\|\Pi\|' = \|\Pi\|$, d'où le Théorème.

3.2. Donnons maintenent une démostration directe du Théorème 3.2, dûe àLax.

Soit $u \in V$, avec $|u| = 1$. Posons: $s_n = (\Pi^n u, \Pi^n u)$, $n=1,2,\ldots,$ $s_o = |u|^2 = 1$. On a (grâce à (3.1)):

$$s_n = (\Pi^n u, \Pi^n u) = (\Pi^{n+1} u, \Pi^{n-1} u)$$

donc

$$s_n^2 \leq s_{n+1}\, s_{n-1}$$

donc

$$s_1 = \frac{s_1}{s_o} \leq \frac{s_2}{s_1} \leq \ldots\ldots\ldots \leq \frac{s_n}{s_{n-1}} \leq \ldots\ldots\ldots$$

d'où

(3.2) $$s_1^n \leqslant s_n .$$

Mais $s_n = |\Pi^n u|^2 \leq c^2 \|\Pi^n u\|^2$ (par (1.1)), donc

(3.3) $$s_n \leq c^2 \|\Pi\|^{2n} \cdot \|u\|^2$$

De (3.2) et (3.3) l'on tire:

$$s_1 \leq c^{2/n} \|u\|^{2/n} \|\Pi\|^2$$

d'où, en faisant augmenter indéfiniment n :

$$s_1 \leq \|\Pi\|^2 \quad \text{, i.e.} \quad |\Pi u| \leq \|\Pi\|$$

si $|u| = 1$, i.e.

$|\Pi u| \leq \|\Pi\| \cdot |u|$ pour tout $u \in V$; d'où le théorème.

4. Opérateur défini par $a(u,v)$

4.1. Soit $a(u,v)$ une forme sesquilinéaire continue sur V, i.e. linéaire en u, anti-linéaire en v, et telle que

$$a(u,v) \leq M \|u\| \|v\| \quad , \quad M = \text{constante}.$$

Si $u \in V$, $v \to a(u,v)$ est donc anti-linéaire continue sur V, donc

$$(4.1) \qquad a(u,v) = (Au,v) \quad , \quad Au \in V' ,$$

et

$$(4.2) \qquad A \in \mathcal{L}(V;V') .$$

Réciproquement, si l'on se donne $A \in \mathcal{L}(V,V')$, alors, $a(u,v)$ définie par (4.1), est sesquilinéaire continue sur V.

4.2- On désigne par $D(A)$ l'espace des $u \in V$ tels que $Au \in H$.

On vérifiera sans peine que la condition nécessaire et suffisante pour qu'un élément u de V soit dans $D(A)$ est que la forme anti-linéaire $v \to a(u,v)$ soit continue sur V pour la topologie induite par H.

$D(A)$ est le domaine de l'opérateur (non borné) A dans H.

4.3. - Dans la suite, nous considérerons une famille d'opérateurs $A(t)$ dépendant du paramètre réel t (le temps), définis dans H à partir de V (ou $V(t)$) et d'une famille de formes $a(t;u,v)$.

5. Une propriété de certains espaces fonctionnels

5.1- De façon générale, si X est un espace de Banach, on désignera par $L^p(\alpha,\beta;X)$ l'espace de Banach des (classes de) fonctions f sur (α,β) mesurables (fortement) à valeurs dans X, telles que

$$\|f\|_{L^p(\alpha,\beta;X)} = \left\{ \int_\alpha^\beta \|f(t)\|_X^p \, dt \right\}^{1/p} < \infty , \quad 1 \leq p < \infty.$$

5.2- <u>On désigne par</u> $W(\alpha, \beta)$ l'espace des (classes de) fonctions u telles que

$$u \in L^2(\alpha, \beta; V) \tag{5.1}$$

$$\frac{du}{dt} \in L^2(\alpha, \beta; V'). \tag{5.2}$$

(Le sens de (5.2) est le suivant: u définit une distribution sur $]\alpha, \beta[$ à valeurs dans V ;alors $\frac{du}{dt}$ est une distribution à valeurs dans V , donc dans V' et (5.2) a un sens).

$W(\alpha, \beta)$ est un <u>espace de Hilbert</u> pour la norme $\|u\|_{W(\alpha,\beta)} = (\|u\|^2_{L^2(\alpha,\beta;V)} + \|u'\|^2_{L^2(\alpha,\beta;V')})^{1/2}$.

Nous voulons démontrer le

<u>Théorème</u> 5.1. <u>Toute fonction</u> u <u>de</u> $W(\alpha, \beta)$ <u>est p.p. égale à une fonction continue de</u> $[\alpha, \beta] \rightarrow H$.

Soit $C([\alpha, \beta]; H)$ l'espace des fonctions continues sur $[\alpha, \beta] \rightarrow H$, muni de la topologie de la convergence uniforme. Alors

$$W(\alpha, \beta) \subset C([\alpha, \beta]; H), \tag{5.3}$$

avec injection continue.

Pour démontrer le théoreme 5.1, commençons par le

<u>Lemme</u> 5.1. <u>Soit</u> $\alpha' < \alpha < \beta < \beta'$. <u>Toute</u> $u \in W(\alpha, \beta)$ <u>est restriction à</u> $[\alpha, \beta]$ <u>de</u> $U \in W(\alpha', \beta')$, <u>telle que</u> U <u>soit nulle dans un voisinage de</u> α' <u>et</u> β' .

<u>Démostration</u>

Supposon (ce qui ne restreint pas la généralité !) que $\alpha = 0$. Définissons v dans $(-\beta, \beta)$ par:

$$v(t) = u(-t) \quad \text{si} \quad t < 0, \qquad u(t) \quad \text{si} \quad t > 0.$$

On vérifie que $v \in W(-\beta,\beta)$. Multipliant v par une fonction θ différentiable, nulle au voisinage de $-\beta$ et égale à 1 au voisinage de $(0,\beta)$, on obtient $w=\theta v$ qui $\in W(-\beta,\beta)$, nulle au voisinage de $-\beta$, et $w=u$ p.p. sur $(0,\beta)$.

Procédé analogue par symétrie autour de β. Le résultat suit.

Lemme 5.2. *Il existe* U_m, *fonctions indéfiniment différentiables dans* $[\alpha',\beta']$ *à valeurs dans* V, *nulles au voisinage de* α' *et* β', *telles que*

$$(5.4.) \qquad U_m \longrightarrow U \quad \text{dans } W(\alpha',\beta')$$

(U *ayant les propriétés énoncées au Lemme* 5.1.).

Démonstration

Par régularisation en t.

Mais pour U_m, on a:

$$|U_m(t)|^2 = \int_{\alpha'}^{t} \frac{d}{d\sigma}(U_m(\sigma), U_m(\sigma))\, d\sigma =$$

$$= \int_{\alpha'}^{t} \Big[(U'_m(\sigma), U_m(\sigma)) + (U_m(\sigma), U'_m(\sigma)) \Big] d\sigma$$

$$< 2 \int_{\alpha}^{t} \| U_m(\sigma) \| \, \| U'_m(\sigma) \|_{V'} \, d\sigma ,$$

d'où

$$(5.5) \qquad |U_m(t)| < C \|U_m\|_{W(\alpha',\beta')} , \quad t \in (\alpha',\beta').$$

De (5.5) résulte (avec (5.4)) que U_m converge uniformément vers U dans H, de sorte que $U \in C([\alpha', \beta'];H)$ (après modification éventuelle sur un ensemble de mesure nulle), d'où résulte le théoreme 5.1 (avec le Lemme 5.1).

5.3. Compléments

Du Lemme 5.2 résulte aussi que l'espace des fonctions indéfiniment différentiables dans $[\alpha, \beta] \longrightarrow V$ est uense dans $W(\alpha, \beta)$.

Par prolengement par continuité, en utilisant le théorème 5.1 , on obtient la formule de Green :

$$\int_{\alpha}^{\beta} \left[(u'(t), v(t)) + (u(t), v'(t))\right] dt = (u(\beta), v(\beta)) - (u(\alpha), v(\alpha)) \tag{5.6}$$

pour $u, v, \in W(\alpha, \beta)$. (Dans cette formule, $u(\alpha)$, $v(\alpha)$, ... désigne la valeur en α de la fonction continue dans H p.p. égale à u, v).

Commentaires sur le Chapitre I

N°1. La condition nécessaire et suffisante pour que $H^m(\Omega) = H_0^m(\Omega)$ est que $\complement \Omega = F$ soit m-polaire, i.e. qu'il n'esiste pas de distribution $T \in H^{-m}(R^n)$ à support dans F qui soit $\neq 0$.

La propriété d'interpolation du résultat du N.3 donnée dans Lions, Bull. Math. R.P.R., Bucarest, 2(1958), p. 419-432, est un cas particulier du résultat suivant (pour lequel nous renvoyons à J.L. Lions - J. Peetre, Espaces de moyennes, à paraitre); soit B un espace de Banach, $B \subset H$, H =Hilbert, B dense dans H, et soit B' l'antidual de B. Donc $B \subset H \subset B'$. Alors si π est un élément de $\mathcal{L}(B';B') \cap \mathcal{L}(B;B)$, c'est aussi un élément de $\mathcal{L}(H;H)$. Cette propriété est inexacte si B n'est pas un Banach - Par exemple si l'on prend $H = L^2(R)$, $B = \mathcal{S}$ = espace de Fréchet des fonctions indéfiniment différentiables à decroissance rapide (cf. L.Schwartz, Théorie des distributions, Paris, Hermann, t. 2), alors $B' = \mathcal{S}'$ = espace des distributions tempérées et $\mathcal{S} \subset L^2 \subset \mathcal{S}'$; les opérateurs de dérivation sont bien linéaires et continus de $\mathcal{S}$ dans lui même et de $\mathcal{S}'$ dans lui même mais n'appliquent pas L^2 dans lui même.

La démonstration du texte du Théorème 3.2 est dûe à P. Lax, On symetrisable transformations, Comm. Pure Applied Maths.

Le N.4 intorduit très brièvement les opérateurs non bornés définis part le couple V, H et une forme sesquilinéaire continue sur V. Pour des applications aux problèmes aux limites "elliptiques", nous renvoyons à J.L.Lions, Problèmes aux limites en théorie des distributions, Acta Math. t. 94 (1955), p.13-153.

Les résultats du N.5 sont des cas particuliers de Lions, Espaces intermédiaires entre espaces hilbertiens et applications, Bull. Math. R. P.R. Bucarest, 2(1958), p.419-432.

Chapitre II

ÉQUATIONS LINÉAIRES DU 1er ORDRE

1. Notations. Exemples

1.1. On désigne par K et H deux espaces de Hilbert, séparables. On suppose que $K \subset H$, l'injection de K dans H étant continue et K dense dans H. Si $f, g \in H$, (f, g) est leur produit scalaire dans H et

$$|f| = (f, f)^{1/2}$$

Si $u, v \in K$, $((u, v))$ est leur produit scalaire dans K et

$$\|u\| = ((u, u))^{1/2} \quad .$$

On considère une variable t - le temps - dans l'intervalle $[0, T]$.

Pour chaque $t \in (0, T)$, on se donne un sous espace vectoriel $V(t)$ fermé dans K, $V(t)$ étant dense dans H.

Le couple $\{V(t), H\}$ jouera le rôle du couple $\{V, H\}$ du Chap. I.

On supposera que $V(t)$ "dépend mesurablement" de t, au sens suivant : désignons par $P(t)$ l'opérateur de projection orthogonale dans K sur $V(t)$.

On supposera alors

$$(1.1) \quad \begin{cases} \text{pour tout } k \in K \text{, la fonction } t \longrightarrow P(t)k \text{ est mesurable sur} \\ [0, T] \text{ à valeurs dans } K . \end{cases}$$

1.2. Exemples.

On prend

$$H = L^2(\Omega) \quad , \quad \Omega \text{ ouvert de } R^n ;$$

$$K = H^m(\Omega) = \left\{ u \;\middle|\; D^p u \in L^2(\Omega) \quad \text{pour tout } p, \; |p| < m \right\} .$$

Muni de la norme

$$\Big(\sum_{|p| \leq m} \int_\Omega | D^p u |^2 dx \Big)^{1/2} = \|u\| ,$$

K est un espace de Hilbert.

L'espace $V(t)$ est défini par des conditions aux limites, définis par des relations linéaires (à coefficients dépendants de t) entre les traces sur la frontière Γ de Ω (Γ est supposée assez régulière) de $u \in H^m(\Omega)$ et des dérivées d'ordre $\leq m-1$.

Sous des hypothèses raisonnables de régularité en t sur les coefficients, on montre alors que les $P(t)$ dépendent mesurablement et même "régulièrement" de t .

1.3. Formes $a(t;u,v)$.

On se donne, pour chaque $t \in (0,T)$, une forme sesquilinéaire continue sur K , avec la propriété:

$$(1.2) \quad \begin{cases} \text{pour tout } u, v \in K \text{ , la fonction } t \to a(t;u,v) \text{ est mesurable} \\ \text{et } |a(t;u,v)| \leq M \|u\| \, \|v\| , \quad M = \text{constante}, \; t \in [0,T] \end{cases}$$

On désigne maintenant par $L^2(0,T;V(t))$ l'espace des (classes de) fonctions u telles que :

i) $u \in L^2(0,T;K)$

ii) $u(t) \in V(t)$ p.p.

On constate facilement que l'on définit ainsi un sous espace vectoriel fermé de $L^2(0,T;K)$; donc $L^2(0,T;V(t))$ est un espace de Hilbert.

Nous aurons besoin du

Lemme 1.1. Si $u, v \in L^2(0,T;V(t))$, la fonction

$$t \longrightarrow a(t;u(t),v(t)) \tag{1.3}$$

est dans $L^1(0,T)$.

Démonstration.

Puisque $|a(t;u(t),v(t))| \leqslant M \|u(t)\| \|v(t)\|$, la seule chose à montrer est la mesurabilité de la fonction (1.3).

Comme $a(t;u,v)$ est continue sur K, on peut écrire

$$a(t;u,v) = ((\mathcal{B}(t)u,v)) \qquad \mathcal{B}(t) \in \mathcal{L}(K;K).$$

et de même, $a(t;u,v)$ étant continue sur $V(t)$:

$$a(t;u,v) = ((\mathcal{A}(t)u,v)), \qquad u,v \in V(t), \qquad \mathcal{A}(t) \in \mathcal{L}(V(t);V(T)).$$

Alors $((\mathcal{B}(t)P(t)u, P(t)v)) = ((\mathcal{A}(t)P(t)u, P(t)v))$ d'où

$$\mathcal{A}(t)P(t) = P(t)\mathcal{B}(t)P(t).$$

On a : $a(t;u(t),v(t)) = ((\mathcal{A}(t)u(t),v(t)))$ et il suffit de montrer que $t \longrightarrow \mathcal{A}(t)u(t)$ est mesurable dans K, donc (K étant séparable) que pour tout $k \in K$,

$$((\mathcal{A}(t)u(t),k)) = ((\mathcal{A}(t)u(t),P(t)k)) = ((u(t),\mathcal{A}^*(t)P(t)k))$$

est mesurable - Donc il suffit de montrer que $t \longrightarrow \mathcal{A}^*(t)P(t)k$ est mesurable donc que $t \longrightarrow ((k_1, \mathcal{A}^*(t)P(t)k))$ est mesurable. Or vaut

$$((k_1, P(t)\mathcal{B}^*(t)P(t)k)) = ((P(t)k_1, \mathcal{B}^*(t)P(t)k)).$$

Donc il suffit de montrer que $t \longrightarrow \mathcal{B}^*(t)P(t)k$ est mesurable, donc que $t \longrightarrow ((\beta^*(t)P(t)k, k_1)) = ((P(t)k, \beta(t)k_1))$ est mesurable. Donc en fin de comptes, il suffit de vérifier que $\beta(t)k_1$ est mesurable, donc que $((\beta(t)k_1, k))$ l'est -

Or vaut $a(t;k_1,k)$ qui est mesurable par hypotehese - c.q.f.d.

2. Le problème. Théorème d'existence

2.1. Nous cherchons $u \in L^2(0,T;V(t))$ telle que

$$(2.1) \qquad \int_0^T \left[a(t;u(t),\varphi(t)) - (u(t),\varphi'(t)) \right] dt = \int_0^T (f(t),\varphi(t))dt + (u_0,\varphi(0))$$

pour toute fonction φ telle que

$$(2.2) \qquad \varphi \in L^2(0,T;V(t)), \quad \varphi' = \frac{d\varphi}{dt} \in L^2(0,T;H), \quad \varphi(T) = 0.$$

Dans (2.1), f est donné dans $(L^2(0,T;V(t)))'$ et u_0 dans H.

Remarque: Grâce au Lemme 1.1, $\int_0^T a(t;u(t),\varphi(t))dt$ a un sens.

2.2. Formellement, la relation (2.1) signifie que

$$(2.3) \qquad \begin{cases} u \in L^2(0,T;V(t)), \\ A(t)u(t) + \dfrac{du}{dt} = f \quad \text{dans } (0,T) \\ u(0) = u_0. \end{cases}$$

En fait le problème 2.1 est une formulation faible (on généralisée) du système (2.3).

On précisera cela completement dans le cas où $V(t) = V$ est indépendant de t.

2.3. Dans la suite nous démontrerons le théorème d'existence:

Théorème 2.1. On suppose que (1.1), (1.2) ont lieu, et qu'il existe une constante λ et une constante $\alpha > 0$ telle que

$$\text{(2.4)} \qquad \operatorname{Re} a(t;v,v) + \lambda |v|^2 \geqslant \alpha \|v\|^2, \quad \text{pour tout} \quad v \in V(t).$$

Alors il existe une fonction u solution du problème 2.1.

N.B. Le problème de l'unicité, sans hypothèses supplémentaires, est ouvert.

Le théorème est démontré au N. suivant.

3. Démonstration du théorème d'existence 2.1.

3.1. Réduction préliminaire :
par changement de u en $e^{kt}u$, k convenable $(> \lambda)$, on peut toujours supposer que (2.4) a lieu avec $\lambda = 0$.

3.2. On va utiliser une méthode de perturbation. On introduit un paramètre $\varepsilon > 0$ destiné à tendre vers 0.

On désigne par W l'espace des (classes de) functions u telles que

$$\text{(3.1.)} \qquad u \in L^2(0,T;V(t)), \qquad u' \in L^2(0,T;H), \qquad u(T) = 0.$$

Muni de la norme

$$\|u\|_W = \left(\int_0^T \left(\|u(t)\|^2 + |u'(t)|^2 \right) dt \right)^{1/2}$$

c'est un espace de Hilbert.

Pour $u, \varphi \in W$, on pose :

$$\text{(3.2)} \qquad b_\varepsilon(u,\varphi) = \int_0^T \left[a(t;u(t), \varphi(t)) - (u(t), \varphi'(t)) + \varepsilon (u'(t), \varphi'(t)) \right] dt.$$

La forme $u, \varphi \longrightarrow b_\varepsilon(u, \varphi)$ est sesquilinéaire continue sur W ; vérifions le :

Lemme 3.1. On a

$$\text{(3.3)} \qquad \operatorname{Re} b_\varepsilon(u,u) \geqslant \alpha \int_0^T \|u(t)\|^2 dt + \varepsilon \int_0^T |u'(t)|^2 dt + \frac{1}{2} |u(0)|^2 .$$

Démonstration

$$2 \operatorname{Re} b_\varepsilon(u,u) = \int_0^T 2 \operatorname{Re} a(t;u,u) dt + 2\varepsilon \int_0^T |u'(t)|^2 dt -$$

$$- \int_0^T \left[(u,u') + (u',u) \right] dt$$

et le dernier terme vaut $- \int_0^T \frac{d}{dt} |u(t)|^2 dt = |u(0)|^2 - |u(T)|^2 = |u(0)|^2$ puisque $u(T) = 0$.

D'où (3.3) (En utilisant (2.4) avec $\lambda = 0$)

Conséquence:

Si $\varphi \longrightarrow L(\varphi)$ est une forme antilinéaire continue sur W, il existe $u_\varepsilon \in W$ unique tel que

$$\text{(3.4)} \qquad b_\varepsilon(u_\varepsilon, \varphi) = L(\varphi) \qquad \text{pour tout } \varphi \in W.$$

En outre, si $|L(\varphi)| \leqslant \|L\| \, \|\varphi\|_W$, on a :

$$\text{(3.5)} \qquad \int_0^T \|u_\varepsilon(t)\|^2 dt + \varepsilon \int_0^T |u'_\varepsilon(t)|^2 dt + |u_\varepsilon(0)|^2 \leqslant \text{constante}.$$

Nous prendrons en particulier :

$$(3.6) \qquad L(\varphi) = \int_0^T (f(t), \varphi(t))\, dt + (u_o, \varphi(0)).$$

D'après (3.5) on peut extraire de u_ε une suite $u_{\varepsilon'}$, $\varepsilon' \to 0$, telle que:

(3.7) $u_{\varepsilon'} \longrightarrow w$ dans $L^2(0,T;V(t))$ faible

(3.8) $\sqrt{\varepsilon'} \frac{d}{dt} u_{\varepsilon'} \to \chi$ dans $L^2(0,T;H)$ faible.

$$\text{Alors} \quad b_{\varepsilon'}(u_{\varepsilon'}, \varphi) \to \int_0^T \left[a(t;w(t), \varphi(t)) - (w(t), \varphi'(t)) \right] dt$$

$$(\text{car} \int_0^T \varepsilon' \left(\frac{du_{\varepsilon'}}{dt}, \frac{d\varphi}{dt}\right) dt \longrightarrow 0)$$

Par conséquent w satisfait à :

$w \in L^2(0,T;V(t))$,

$$\int_0^T \left[a(t;w(t), \varphi(t)) - (w(t), \varphi'(t)) \right] dt = \int_0^T (f(t), \varphi(t))\, dt + (u_o, \varphi(0))$$

et on peut donc prendre $u = w$, ce qui achève la démonstration du Théorème.

3.3. Remarque.

Le problème "perturbé" (3.4), avec le choix (3.6) de $L(\varphi)$, correspond formellement à :

$$A(t)\, u_\varepsilon(t) + u'_\varepsilon(t) - \varepsilon u''_\varepsilon(t) = f,$$

$$u_\varepsilon(0) - \varepsilon u'_\varepsilon(0) = u_o,$$

$$u_\varepsilon(T) = 0.$$

4. Le cas $V(t) = V$.

4.1. Si $V(t) = V$ indépendant de t, on peut dans (2.1) prendre les fonctions φ de la façon suivante :

$$(4.1)\quad \begin{cases} \varphi(t) = \psi(t)v, \\ v \in V,\ \psi \in L^2(0,T),\ \psi' \in L^2(0,T),\ \psi(T) = 0. \end{cases}$$

Alors (2.1) s'écrit :

$$(4.2)\quad \begin{cases} \int_0^T a(t;u(t),v)\,\overline{\psi(t)}\,dt - \int_0^T (u(t),v)\,\overline{\psi'(t)}\,dt = \\ = \int_0^T (f(t),v)\,\overline{\psi(t)}\,dt + (u_0,v)\,\overline{\psi(0)}. \end{cases}$$

Ici $f \in L^2(0,T;V')$ $(=(L^2(0,T;V))')$.

Si en particulier nous prenons

$\psi \in \mathscr{D}(]0,T[)$ (fonction indéfiniment différentiable à support compact dans $(0,T)$), alors (4.2) se réduit à

$$\int_0^T \left(a(t;u(t),v) + \frac{d}{dt}(u(t),v)\right)\overline{\psi(t)}\,dt = \int_0^T (f(t),v)\,\overline{\psi(t)}\,dt$$

pour tout $\psi \in \mathscr{D}(]0,T[)$, où $\frac{d}{dt}(u(t),v)$ est prise au sens des distributions dans $]0,T[$. Donc :

$$(4.3)\quad a(t;u(t),v) + \frac{d}{dt}(u(t),v) = (f(t),v) \quad \underline{\text{au sens de}}\ \mathscr{D}'(]0,T[)$$

$(\mathscr{D}'(]0,T[) = \mathscr{D}(]0,T[)' =$ distributions sur $]0,T[$).

Mais(cf. Chap.I) :

(4.4) $\quad a(t;u(t),v) = (A(t)\,u\,(t),v)\ , \qquad A(t)\,u\,(t) \in V'$.

Notons le :

<u>Lemme 4.1.</u> $A(t)\,u\,(t) \in L^2(0,T;V')$.

<u>Démonstration.</u> Il suffit de vérifier la <u>mesurabilité</u> de cette fonction; donc que, pour $v \in V, t \longrightarrow (A(t)\,u\,(t), v)$ est mesurable. Donc que $t \longrightarrow (u(t), A^*(t)v)$ l'est; donc que $t \longrightarrow A^*(t)v$ est mesurable, donc que $t \longrightarrow (w, A^*(t)v)$ l'est, donc que $t \longrightarrow (A(t)w, v) = a(t;w,v)$ l'est, ce qui est vrai par hypothèse. C.q.f.d.

Alors (4.3) donne:

(4.5) $$A(t)\,u\,(t) + \frac{du(t)}{dt} = f(t)$$

et par conséquent

$$\frac{du(t)}{dt} = f(t) - A(t)\,u\,(t) \in L^2(0,T;V').$$

Donc :

<u>Théorème</u> 4.1. <u>Si</u> $V(t) = V$ <u>indépendant de</u> t <u>, et si</u> u <u>est solution du Problème</u> 3.1, <u>alors</u>

(4.6) $$\frac{du}{dt} \in L^2(\ 0,T;V').$$

<u>Corollaire</u> 4.1. <u>Sous les hypothèses du Théorème</u> 4.1, u <u>est (p.p. égale à une fonction) continue de</u> $[0,T] \longrightarrow H$ <u>et</u> $u(0) = u_0$.

<u>Démonstration.</u>

La 1ère partie du théorème est conséquence du Chap. I et (4.6).

En outre, on peut écrire maintenant (cf. Chap. I) que

$$-\int_0^T (u(t), \varphi'(t))\,dt = \int_0^T (u'(t), \varphi\,(t))\,dt + (u(0), \varphi\,(0))$$

(puisque φ (T) = 0). Donc :

$$\int_0^T \left[a(t;u(t), \varphi(t)) + (u'(t), \varphi(t))\right] dt + (u(0), \varphi(0)) = \int_0^T (f(t), \varphi(t)) dt + (u_0, \varphi(0))$$

et d'aprés (4.5):

$$\int_0^T \left[a(t;u(t), \varphi(t)) + (u'(t), \varphi(t))\right] dt = \int_0^T (f(t), \varphi(t)) dt$$

donc $(u(0), \varphi(0))$ $(u_0, \varphi(0))$ donc $u(0) = u_0$ ce qui achève la démonstration du corollaire.

4.2. Unicité

Théorème 4.2. Hypothèses du Théorème 2.1 avec $V(t) = V$. Alors le problème 2.1 admet une solution unique.

Démonstration.

D'après le point 4.1, il faut démontrer ceci : si u vérifie

$$u \in L^2(0,T;V) \ , \quad \frac{du}{dt} \in L^2(0;T;V'), \tag{4.7}$$

$$A(t)\, u + \frac{du}{dt} = 0, \tag{4.8}$$

$$u(0) = 0, \tag{4.9}$$

alors $u \equiv 0$. Or de (4.8) résulte :

$$\int_0^T a(t;u(t), u(t))\, dt + \int_0^T \left(\frac{du(t)}{dt}, u(t)\right) dt = 0, \tag{4.10}$$

Mais (cf. Chap. I)

$$\int_0^T \left[\left(\frac{du(t)}{dt}, u(t)\right) + \left(u(t), \frac{du(t)}{dt}\right)\right] dt = |u(T)|^2 - |u(0)|^2 = |u(T)|^2$$

donc prenant la partie réelle de (4.10), il vient

$$\int_0^T \text{Re } a(t;u(t),u(t))\, dt + \frac{1}{2} |u(T)|^2 = 0$$

et comme $\text{Re } a(t;u(t),u(t)) \geq \alpha \| u(t)\|^2$, $\alpha > 0$, on en déduit que $u \equiv 0$. c.q.f.d.

5. Une application de la méthode de transposition

5.1. On peut énoncer le résultat final du N.4 (en prenant $u_o = 0$) sous la forme suivante :

$$u \longrightarrow A(t)u + \frac{du}{dt}$$

est un isomorphisme de l'espace X(des $u \in L^2(0,T;V)$ tels que $u' \in L^2(0,T;V')$, $u(0) = 0$) sur l'espace $L^2(0,T;V')$.

En "inversant le sens du temps" et changeant $A(t)$ en $A^*(t)$, alors

$$v \longrightarrow \Sigma v = A^*(t)\, v - \frac{dv}{dt}$$

est un isomorphisme de l'espace X_+ (des $v \in L^2(0,T;V)$ tels que $v' \in L^2(0,T;V')$, $v(T) = 0$) sur l'espace $L^2(0,T;V')$.

5.2. On transpose maintenant ce résultat : Σ^*(adjoint de Σ) est un isomorphisme de $L^2(0,T;V)$ sur $(X_+)'$. Donc :

$$(5.1)\quad \begin{cases} \text{étant donnée } v \longrightarrow L(v) \text{, forme anti-linéaire continue sur} \\ X_+ \text{, il existe } u \in L^2(0,T;V) \text{ unique tel que} \\ (u, \Sigma v) = L(v) \text{ pour tout } v \in X_+ . \end{cases}$$

$(u, \Sigma v)$ désigne ici le produit scalaire entre $L^2(0,T;V)$ et $L^2(0,T;V')$-

5.3. Cas particulier.

Soit $f \in L^1(0,T;H)$ (L^1 et non L^2 !) donnée et soit

$$\text{(5.2.)} \quad L(v) = \int_0^T (f(t), v(t))\,dt + (u_0, v(0)).$$

Comme (cf. Chap. I) $X_+ \subset C(0,T;V)$ (fonctions continues de $0,T \longrightarrow H$), $L(v)$ définie par (5.2.) est bien un élement de $(X_+)'$ (N.B. : on pourrait même prendre pour f une mesure à valeurs dans H).

Donc :

(5.3)
- il existe $u \in L^2(0,T;V)$ unique, telle que
- $$\int_0^T (u(t), A^*(t)v(t) - v'(t))\,dt = \int_0^T (f(t), v(t))\,dt + (u_0, v(0))$$
- où $f \in L^1(0,T;H)$, $u_0 \in H$, pour tout $v \in L^2(0,T;V)$ tel que $v' \in L^2(0,T;V')$, $v(T) = 0$.

On déduit de (5.3) que

$$\text{(5.4)} \quad A(t)\,u(t) + u'(t) = f(t) \quad \text{dans }]0,T[$$

Donc, cette fois :

$$\text{(5.5)} \quad u' \in L^2(0,T;V') + L^1(0,T;H).$$

On peut montrer que dans ces conditions u est encore (p.p. égale à) une fonction continue à valeurs dans H. Alors

$$\text{(5.5)} \quad u(0) = u_0 .$$

Commentaires sur le Chap. II

Le problème de l'unicité dans le Théorème 2.1 n'est pas résolu.

Par ex. prenons; $H = L^2(\Omega)$, $K = H^1(\Omega)$, Ω ouvert borné de frontière assez régulière, et soit $\sum(t)$ une partie de Γ = frontière de Ω dépendant de t, par ex. continûment. Prenons :

$$V(t) = \left\{ u \mid u \in K = H^1(\Omega), \quad u = 0 \text{ sur } \sum(t) \right\}$$

$$a(t;u,v) = a(u,v) = \sum_{i=1}^{n} \int_{\Omega} \frac{\partial u}{\partial x_i} \overline{\frac{\partial v}{\partial x_i}} \, dx.$$

Le problème correspondant au Problème 2.1 n'est pas résolu, à notre connaissance (sauf, évidemment, si $\sum(t)$ ne dépend pas de t !).

Il correspond, formellement, au problème suivant :

$$-\Delta_x u(x,t) + \frac{\partial u}{\partial t} = f \qquad \text{dans } \Omega \times \left]0, T\right[,$$

$u(x, 0)$ donné

$u(x,t) = 0$ si $x \in \sum(t)$, $t \in (0,T)$

$\frac{\partial u}{\partial n}(x,t) = 0$ si $x \in \Gamma - \sum(t)$, $t \in (0,T)$.

Si l'on impose des hypothèses supplémentaires, alors des résultats d'unicité sont connus :

1) Si l'on suppose que $a(t;u,v)$ dépend de façon différentiable de t, et que $P(t)$ dépend aussi de façon différentiable de t, alors l'unicité est démontrée dans

T. Kato - H. Tanabe, On the abstract evolution equation, Osaka Math. J. 14 (1962), p. 107-133.

2) L'hypothèse de différentiabilité sur $a(t;u,v)$ (mais pas sur $P(t)$)

a été supprimée dans

J.L. Lions, Remarques sur les équations différentielles opérationelles, à paraître.

On trouvera une démonstration du Théorème 2.1 différente de celle donnée au N. 3 dans

J.L. Lions, Springer Collection Jaune, t. 111, 1961.

Le N. 5 est plus important pour la méthode que pour le résultat; cette méthode de transposition, inaugurée par Visik et Sobolev, est essentielle dans les problèmes aux limites non homogènes étudiés par Magenes et Lions - et aussi Schechter - Pour le cas présent, on pourra consulter pour quelques compléments :

J.L. Lions, Quelques remarques sur les équations différentielles opérationelles, Rend. di Padova, 1963.

J. L. Lions

Chapitre III

INTRODUCTION À CERTAINS PROBLÈMES NON LINÉAIRES

1. Position du problème. Généralités.

Nous voulons résoudre le problème de Cauchy pour l'équation

(1.1) $$\Lambda u = \frac{\partial u}{\partial t} - \sum_{i=1}^{n} \frac{\partial}{\partial x_i} \left(\frac{\partial u}{\partial x_i}\right)^{2k-1} = f,$$

k entier ≥ 2.

Donc on cherche une "solution" (dans un sens ou dans un autre !) de (1.1) définie pour $t > 0$ et $x \in R^n$, telle que

(1.2) $$u(x, 0) = u_o(x), \quad u_o \text{ donnée,}$$

$u(x, t)$ étant assujettie, pour chaque $t \geqslant 0$, à une condition de croissance à l'infini (en x).

Toutes les fonctions considérées, dans ce chapitre sont à valeurs réelles.

Pour les problèmes non linéaires d'évolution (et même pour les problèmes linéaires - comme on verra -) une méthode très puissante est la méthode des solutions approchées.

Expliquons cette méthode en l'adaptant au cas du problème ci dessus; on va d'ailleurs voir que dans ce cas elle est insuffisante.....

On prend une suit de fonctions

$$w_1, w_2, \ldots\ldots, w_m, \ldots\ldots$$

telles que

$w_i \in \mathcal{D}(R^n)$ pour tout i,
quel que soit m, les $w_1 \ldots \ldots w_m$ sont linéairement indépendentes, les combinaisons linéaires $\sum_{\text{finie}} \xi_i w_i$, $\xi_i \in R$, sont denses dans (R^n).

<u>On peut remplacer</u> $\mathcal{D}(R^n)$ <u>par n'importe quel espace de fonctions régulières à décroissance suffisamment rapide à l'infini.</u>

On cherche alors une "solution approchée" u_m sous la forme

$$u_m(t) = \sum_{i=1}^{m} g_{im}(t)\, w_i \tag{1.3}$$

(ou encore : $u_m(x, t) = \sum_{i=1}^{m} g_{im}(t)\, w_i(x)$),

où les $g_{im}(t)$ sont déterminées de façon à satisfaire à :

$$\left(\frac{du_m(t)}{dt}, w_j\right) + \sum_{i=1}^{n} \left(\left(\frac{\partial u_m}{\partial x_i}\right)^{2k-1}, \frac{\partial w_i}{\partial x}\right) = 0, \; j = 1, \ldots m \tag{1.4}$$

où de façon générale

$$(f, g) = \int_{R^n} f(x) g(x)\, dx;$$

on ajoute à (1.4) les conditions initiales :

$$g_{im}(0) = \alpha_{im}, \tag{1.5}$$

les α_{im} étant choisis de façon que $\sum_{i=1}^{m} \alpha_{im} w_i \longrightarrow u_o$, la convergence ayant lieu dans un espace convenable (on précisera cela dans la suite !).

Le système (1.4) (1.5) <u>est un système d'équations différentielles ordinaires non linéaires</u>.

Montrons que (1.4) (1.5) définissent globalement (i.e. pour <u>tout</u> $t \geq 0$) les $g_{im}(t)$. Pour cela, multiplions (1.4) par $g_{im}(t)$ et sommons en j ; nous obtenons :

$$\left(\frac{du_m(t)}{dt}, u_m(t)\right) + \sum_{i=1}^{n} \left(\left(\frac{\partial u_m}{\partial x_i}\right)^{2k-1}, \frac{\partial u_m}{\partial x_i}\right) = 0 ,$$

ou encore

$$\frac{1}{2}\frac{d}{dt} | u_m(t)|^2 + \sum_{i=1}^{n} \int_{R^n} \left(\frac{\partial u_m}{\partial x_i}\right)^{2k} dx = 0 \quad (| \ | = \text{norme dans } L^2(R^n))$$

et donc

$$(1.6) \qquad |u_m(t)|^2 + 2 \sum_{i=1}^{n} \int_0^t \int_{R^n} \left(\frac{\partial u_m}{\partial x_i}\right)^{2k} dx.dt = |u_m(0)|^2$$

Supposons alors que $u_o \in L^2(R^n)$ et que $\sum_{i=1}^{m} \alpha_{im} w_i \rightarrow u_o$ dans $L^2(R^n)$.

De (1.6) résulte alors que, pour $o \leq t \leq T$, T fini quelconque

$$(1.7) \qquad \begin{cases} | u_m(t)| \leq \text{constante} \\ \dfrac{\partial u_m}{\partial x_i} \text{ borné dans } L^{2k}(R^n x(0, T)) \text{ lorsque } m \text{ varie (et} \end{cases}$$

de tout ceci résulte que $u_m(t)$ est défini pour <u>tout</u> $t \geq 0$).

Le problème est maintenant <u>de passer à la limite,</u> en extrayant (si possible!) de u_m une suite u_ν telle que u_ν converge vers une solution du problème initial.

Or, d'après (1.7), on peut extraire une suite u_ν telle que

(1.8) $u_\nu \longrightarrow w$ dans $L^\infty(0,T;L^2(R^n))$ faible,

(1.9) $\frac{\partial u_\nu}{\partial x_i} \longrightarrow \chi_i$ dans $L^{2k}(R^n x(0,T))$ faible.

On vérifie sans peine que

(1.10) $$\chi_i = \frac{\partial w}{\partial x_i}$$

La difficulté principale (la seule même!) est de montrer que $(\frac{\partial u_\nu}{\partial x_i})^{2K-1}$ converge, même dans un sens tres faible, vers $(\frac{\partial w}{\partial x_i})^{2K-1}$.

Par exemple, puisque d'après (1.7), $(\frac{\partial u_m}{\partial x_i})^{2K-1}$ est borné dans $L^{\frac{2K}{(2K-1)}}(R^n x(0,T))$ on peut supposer que

$$(\frac{\partial u_\nu}{\partial x_i})^{2K-1} \longrightarrow \psi_i \quad \text{dans} \quad L^{\frac{2K}{(2K-1)}}(R^n x(0,T)) \text{ faible}$$

mais, à cause de la convergence faible, on ne peut en conclure que

$$\psi_i = (\frac{\partial w}{\partial x_i})^{2K-1} .$$

Autrement dit, les inégalités à priori fournies par (1.6) ne sont pas suffisantes; le problème est d'obtenir des estimations à priori portant sur $\frac{\partial^2 u}{\partial x_i \partial t}$, $\frac{\partial^2 u_m}{\partial x_j \partial x_i}$.

2) Les estimations à priori fondamentales. Applications.

2.1. On va utiliser l'opérateur auxiliaire B donné par

(2.1à) $$Bu = -\Delta u + u - \frac{\partial}{\partial t}(T-t)\frac{\partial u}{\partial t} .$$

On simplifie un peu en prenant $u_0 = 0$ dans le N. 1 - et, la base w_j étant choisie comme au N. 1, on cherche à déterminer $u_m(t)$ par :

(2.2) $$u_m(t) = \sum_{i=1}^{m} g_{im}(t)\, w_i ,$$

avec :

(2.3) $$(B \Lambda\, u_m(t), w_j) = (B f, w_j) , \quad j=1, \ldots, m,$$

(ce qui suppose f assez régulière) avec les conditions

(2.4) $$u_m(0) = 0, \quad u'_m(0) = 0, \quad u_m(t) \text{ borné lorsque } t \longrightarrow T.$$

Nous admettons ici le lemme suivant (pour le quel nous renvoyons à Visik, M. Visik utilise ici la méthode de Leray-Schauder)

Lemme 2.1. On suppose que f est indéfiniment différentiable à support compact dans l'ouvert $t > 0$. Il existe alors une fonction u_m satisfaisant à (2.2), (2.3), (2.4).

Nous allons alors démontrer la

Proposition 2.1. Hypothèse du Lemme 2.1 sur f. Il existe une fonction u telle que

(2.5) $$u \in L^{\infty}(0, T; L^2(R^n)) ,$$

(2.6) $$\frac{\partial u}{\partial x_i} \in L^{2k}(R^n x (0, T)) ,$$

(2.7) $$\Lambda\, u = \frac{\partial u}{\partial t} - \sum_{i=1}^{n} \frac{\partial}{\partial x_i}\left(\frac{\partial u}{\partial x_i}\right)^{2k-1} = f \quad \text{dans} \quad t > 0 ,$$

(2.8) $$u(x, 0) = 0.$$

Remarque 2.1. La condition (2.8) a un sens. En effet, posons :

$$2K = P$$

Alors $\left(\frac{\partial u}{\partial x_i}\right)^{2K-1} \in L^{\frac{2K}{2K-1}}(R^n x (0,T)) = L^{p'}(R^n \times (0,T))$

où $\frac{1}{p} + \frac{1}{p'} = 1$. Donc, en séparant les variables :

$$\left(\frac{\partial u}{\partial x_i}\right)^{2K-1} \in L^{p'}(0,T;L^{p'}(R^n)). \tag{2.9}$$

Soit $W^{1,p}(R^n) = \left\{\varphi \,\middle|\, \varphi \in L^p,\ \frac{\partial \varphi}{\partial x_i} \in L^p(R^n),\ i = 1,\ldots,n\right\}$

et soit $W^{-1,p'}(R^n)$ = dual de $W^{1,p}(R^n)$. Alors $\frac{\partial}{\partial x_i}$ est un opérateur linéaire continu de $L^{p'}(R^n) \longrightarrow W^{-1,p'}(R^n)$ et donc (2,9) entraine :

$$\frac{\partial}{\partial x_i}\left(\frac{\partial u}{\partial x_i}\right)^{2K-1} \in L^{p'}(0,T;W^{-1,p'}(R^n)). \tag{2.10}$$

De (2.7), (2.10) l'on déduit

$$\frac{\partial u}{\partial t} \in L^{p'}(0,T;W^{-1,p'}(R^n)). \tag{2.11}$$

Mais (2.11) implique en particulier que u est (p.p. égale à une fonction) continue de $[0,T] \rightarrow W^{-1,p'}(R^n)$ (on peut préciser : de (2.5) et (2.11) résulte que u est continue de $[0,T]$ à valeurs dans un espace intermédiaire entre $L^2(R^n)$ et $W^{-1,p'}(R^n)$) . Donc (2.8) a un sens.

Avant de démontrer la Proposition 2.1 établissons les inégalités à priori :

2.2. Les inégalités à priori.

Nous multiplions (2.3) par $g_{im}(t)$ et sommons en j . Il vient, en intégrant en t sur $(0,T)$:

(2.12) $$J_1 + J_2 = \int_o^T (Bf,\ u_m(t))\,dt\ ,$$

où

$$J_1 = - \int_o^T \left(\frac{\partial}{\partial t}(T-t)\frac{\partial}{\partial t}\Lambda u_m, u_m\right) dt\ ,$$

$$J_2 = \int_o^T (-\Delta\Lambda u_m + \Lambda u_m, u_m)\,dt\ .$$

Mais

$$J_1 = \int_o^T (T-t)\left(\frac{\partial}{\partial t}\Lambda u_m, \frac{\partial u_m}{\partial t}\right) dt$$

$$= \int_o^T (T-t)\,(u''_m(t), u'_m(t))\,dt + \qquad \left\{\text{on note } u' = \frac{\partial u}{\partial t}\right\}$$

$$\underbrace{+ \sum_{i=1}^n \int_o^T (T-t)\left(\int_{R^n} \frac{\partial}{\partial t}\left(\frac{\partial u_m}{\partial x_i}\right)^{2k-1} \frac{\partial^2 u}{\partial x_i \partial t}\,dx\right) dt}_{J_1^{(i)}}$$

Soit $J_{1,o}$ le premier terme et $J_1^{(i)}$ les autres, de sorte que

$$J_1 = J_{1,o} + \sum_{i=1}^{n} J_1^{(i)} .$$

Le premier vaut

$$J_{1,o} = \frac{1}{2} \int_o^T \left| u'_m(t) \right|^2 dt . \tag{2.13}$$

On peut écrire $J_1^{(i)}$ sous la forme :

$$J_1^{(i)} = (2k-1) \int_o^T (T-t)\, dt \int_{R^n} \left(\frac{\partial u_m}{\partial x_i}\right)^{2k-2} \left(\frac{\partial^2 u_m}{\partial x_i \partial t}\right)^2 dx .$$

Introduisons la fonction

$$\Phi(\lambda) = \begin{cases} \lambda^k \text{ si } k \text{ impair} \\ \lambda^k (\text{Signe } \lambda) \text{ si } k \text{ pair.} \end{cases} \tag{2.14}$$

Alors (comme $\frac{\partial u_m}{\partial x_i} = o$ là où elle change de signe) :

$$\left(\frac{\partial u_m}{\partial x_i}\right)^{2k-2} \left(\frac{\partial^2 u_m}{\partial x_i \partial t}\right)^2 = \frac{1}{k^2} \left[\frac{\partial}{\partial t} \Phi\left(\frac{\partial u_m}{\partial x_i}\right)\right]^2 .$$

Donc :

$$J_1^{(i)} = \frac{2k-1}{k^2} \int_o^T \int_{R^n} (T-t) \left[\frac{\partial}{\partial t} \Phi\left(\frac{\partial u_m}{\partial x_i}\right)\right]^2 dx\, dt.. \tag{2.15}$$

Calculons maintenant J_2 ; posons :

$$((u,v)) = \int_{R^n} \left(uv + \sum_{i=1}^{n} \frac{\partial u}{\partial x_i} \frac{\partial v}{\partial x_i}\right) dx .$$

Alors

$$J_2 = \underbrace{\int_o^T ((u'_m(t), u_m(t))) \, dt}_{J_{2,o}} + \underbrace{\int_o^T \int_\Omega \left(\frac{\partial u_m}{\partial x_i}\right)^{2k} dx \, dt}_{J'_{2,o}} + \sum_{i=1}^n J_2^{(i)}$$

où

$$J_2^{(i)} = \int_o^T dt \int_{R^n} -\Delta\left(\left(\frac{\partial u_m}{\partial x_i}\right)^{2k-1}\right) \frac{\partial u_m}{\partial x_i} \, dx .$$

Mais

$$J_2^{(i)} = \int_o^T dt \sum_{j=1}^n \int_{R^n} \frac{\partial}{\partial x_j}\left(\left(\frac{\partial u_m}{\partial x_i}\right)^{2k-1}\right) \frac{\partial^2 u_m}{\partial x_i \partial x_j} \, dx$$

et par une trasformation analogue à celle effectuée pour $J_1^{(i)}$:

$$(2.16) \qquad J_2^{(i)} = \frac{2k-1}{k^2} \sum_{j=1}^n \int_o^T \int_{R^n} \left[\frac{\partial}{\partial x_j} \Phi\left(\frac{\partial u_m}{\partial x_i}\right)\right]^2 dx \, dt.$$

Par ailleurs

$$(2.17) \qquad J_{2,o} = \frac{1}{2} \int_o^T \frac{d}{dt} \| u_m(t) \|^2 \, dt = \frac{1}{2} \| u_m(T) \|^2.$$

Donc :

$$
(2.18) \quad \left\{ \begin{array}{l} J_{1,o} + \sum_{i=1}^{n} (J_1^{(i)} + J_2^{(i)}) + J'_{2,o} + \frac{1}{2} \| u_m(T) \|^2 = \\ \qquad\qquad = \int_o^T (Bf, u_m(t))dt, \\ J_{1,o} \text{ donné par (2.13),} \\ J_1^{(i)} \text{ (resp. } J_2^{(i)} \text{) donné par (2.15) (resp. (2.16)),} \\ J'_{2,o} = \int_o^T \int_\Omega \left(\frac{\partial u_m}{\partial x_i}\right)^{2k} dx\, dt. \end{array} \right.
$$

On déduit de là que :

(2.19) $\quad J_{1,o} + \sum_{i=1}^{n} (J_1^{(i)} + J_2^{(i)}) + J'_{2,o}$ <u>est borné lorsque</u> $m \longrightarrow +\infty$

2.3. - <u>Utilisation des inégalités à priori</u> - <u>Démonstration de la Proposition</u> 2.1.

De (2.19) l'on déduit que l'on peut extraire de u_m une suite u_ν telle que :

(2.20)
$$\begin{cases} u_\nu \longrightarrow w \quad \text{dans} \quad L^\infty(0,T;L^2(R^n)) \quad \text{faible}, \\ u'_\nu = \dfrac{\partial u_\nu}{\partial t} \longrightarrow \dfrac{\partial w}{\partial t} \quad \text{dans} \quad L^2(0,T;L^2(R^n)) \quad \text{faible}, \\ \dfrac{\partial u_\nu}{\partial x_i} \longrightarrow \dfrac{\partial w}{\partial x_i} \quad \text{dans} \quad L^{2k}(R^n \times (0,T)) \quad \text{faible}, \\ \left(\dfrac{\partial u_\nu}{\partial x_i}\right)^{2k-1} \longrightarrow \chi_i \quad \text{dans} \quad L^{2k/(2k-1)}(R^n \times (0,T)) \quad \text{faible}, \end{cases}$$

(2.21) $$\Phi\left(\frac{\partial u_\nu}{\partial x_i}\right) \quad \text{converge p.p.}$$

Seul (2.21) doit être démontré : d'après l'expression de $J_1^{(i)} + J_2^{(i)}$, $\Phi\left(\frac{\partial u_m}{\partial x_i}\right)$ demeure dans un borné de $H^1(R^n x(0, T-\varepsilon))$, $\varepsilon > 0$ quelconque; on peut donc extraire une suite u_ν telle que $\Phi\left(\frac{\partial u_\nu}{\partial x_i}\right)$ converge fortement dans L^2 sur tout compact de $R^n x(0,T)$ et donc, par une nouvelle extraction, on peut supposer que (2.21) a lieu.

Mais (cf. (2.14)) la fonction $\Phi(\lambda)$ est monotone, de sorte que (2.21) implique :

(2.22) $$\frac{\partial u_\nu}{\partial x_i} \quad \text{converge p.p.}$$

Mais de (2.22) et du fait que $\frac{\partial u_\nu}{\partial x_i}$ est borné dans $L^{2k}(R^n x(0,T))$, résulte en T que, si θ_i est la limite p.p. de $\frac{\partial u_\nu}{\partial x_i}$, $\frac{\partial u_\nu}{\partial x_i} \to \theta_i$ au sens des distributions; donc $\theta_i = \frac{\partial w}{\partial x_i}$ et donc (2.22) peut se préciser:

(2.23) $$\frac{\partial u_\nu}{\partial x_i} \longrightarrow \frac{\partial w}{\partial x_i} \quad \text{p.p.}$$

De là on déduit que (en utilisant (2.20)) :

(2.24) $$\left(\frac{\partial u_\nu}{\partial x_i}\right)^{2k-1} \longrightarrow \left(\frac{\partial w}{\partial x_i}\right)^{2k-1} \quad \text{dans } L^{2k/(2k-1)}(R^n x(0,T)) \text{ faible.}$$

Passons maintenant à la

Démonstration de la Proposition 2.1

Si φ_j est une fonction scalaire, par exemple dans $\mathcal{D}(]0,T[)$, on multiplie (2.3) (écrite pour $m=\nu$) par φ_j et l'on somme en j, $j \leq m_o \leq m$.

Si l'on pose :

(2.25) $$\psi(t) = \sum_{j=1}^{m_o} \varphi_j(t)\, w_j$$

il vient :

$$(B \Lambda u_m(t), \psi(t)) = (B f, \psi(t)) \quad , \quad (m = \nu)$$

d'où en intégrant sur $(0,T)$ puis par parties :

$$\int_o^T \int_{R^n} \Lambda u_m B \psi \, dx\, dt = \int_o^T \int_{R^n} f B \psi \, dx\, dt ,$$

ce que l'on peut encore écrire ;

$$\left.\begin{aligned} &\int_o^T (u'_m, B\psi(t))\, dt + \sum_{i=1}^n \int_o^T \int_{R^n} \left(\frac{\partial u_m}{\partial x_i}\right)^{2k-1} \frac{\partial}{\partial x_i} B \psi \, dx\, dt = \\ &\qquad = \int_o^T \int_{R^n} f B \psi \, dx\, dt \end{aligned}\right\} \; m = \nu$$

et grâce à (2.20), (2.24) on peut passer à la limite; il vient :

(2.26)
$$\left[\begin{aligned}&\int_0^T (w'(t), B\varphi(t))\,dt + \sum_{i=1}^n \int_0^T \int_{R^n} \left(\frac{\partial w}{\partial x_i}\right)^{2k-1} \left(\frac{\partial}{\partial x_i} B\varphi\right) dx\,dt \\ &= \int_0^T \int_{R^n} f\, B\varphi\, dx\,dt\,.\end{aligned}\right.$$

Cette relatior a lieu puor toute φ de la forme (2.25).

Enfait tout va bien aussi si l'on prend $\varphi_j(t)$ satisfaisant à :

(2.27)
$$\left\{\begin{aligned}&\varphi_j \text{ est deux fois continûment différentiable dans } [0,T[\,,\\ &\varphi_j(0) = 0,\quad \varphi_j(t) \text{ bornée lorsque } t \to T.\end{aligned}\right.$$

Admettons un istant le

<u>Lemme</u> 2.2. <u>On peut choisir une base</u> w_j <u>telle que les fonctions</u>

$$\varphi = \sum_{\text{finie}}' \varphi_j(t)\, w_j\,, \qquad \varphi_j \text{ satisfaisant à (2.27)}$$

<u>soient telles que les</u> $B\varphi$ <u>forment un système dense dans</u>

$L^p(0,T;W^{1,p}(R^n)) \cap L^2(R^n x(0,T))$, $p = 2k$.

Alors (2.26), qui peut s'écrire :

$$\left\langle \frac{\partial w}{\partial t} - \sum_{i=1}^n \frac{\partial}{\partial x_i}\left(\frac{\partial w}{\partial x_i}\right)^{2k-1} - f\,,\ B\varphi \right\rangle = 0$$

(dualité entre L^2 et lui même et $L^{p'}(W^{-1,p'})$ et $L^p(W^{1,p})$ -cf. Remarque 2.1), entraine :

$$\frac{\partial w}{\partial t} - \sum_{i=1}^{n} \frac{\partial}{\partial x_i} \left(\frac{\partial w}{\partial x_i}\right)^{2k-1} = f.$$

On peut alors prendre $u = w$ et la Proposition 2.1 est démontrée, sous réserve de la

Démonstration du Lemme 2.2.

Soient $\theta_k(t)$ les fonctions propres de $-\frac{d}{dt}(T-t)\frac{d}{dt}$:

$$\begin{cases} -\frac{d}{dt}(T-t)\frac{d}{dt} \quad \theta_k(t) = \mu_k \theta_k, \quad k=1,\ldots, \quad \mu_k > 0 \\ \theta_k(0) = 0, \quad \theta_k(t) \text{ bornée lorsque } t \longrightarrow T. \end{cases}$$

Soit ensuite $v_{km}(x)$ vérifiant

$$-\Delta v_{km} + (1+\mu_k) v_{km} = \xi_m \quad , \quad m = 1, \ldots.$$

les ξ_m, $m = 1, \ldots,$ formant une base de $\mathcal{D}(R^n)$, avec, par ex., $v_{km} \in L^2(R^n)$.

Alors, les ξ_{km} étant des nombres réels quelconques :

$$B\left(\sum_{k,m} \xi_{km} v_{km}(x)\, \theta_k(t)\right) = \sum_{k,m}' \xi_{km}\, \xi_m(x)\, \theta_k(t) \quad ,$$

(k, m variant dans en ensemble fini). En prenant les w_i égaux aux v_{km}, on obtient le résultat désiré. Les w_i ne sont pas dans $\mathcal{D}(R^n)$ mais à décroissance à l'infini suffisante pour valider tout ce qui précède.

3. Le résultat final. Théorème d'existence et d'unicité.

3.1. Enoncé du résultat.

On pose : $2k = p$; p' est défini par $1/p + 1/p' = 1$

Théorème 3.1. On donne f avec

$$f \in L^{p'}(o, T; W^{-1,p'}(R^n)) \tag{3.1.}$$

Il existe une fonction u et une seule telle que

$$\begin{cases} u \in L^{\infty}(0, T; L^2(R^n)) \\ u \in L^p(0, T; W^{1,p}(R^n)) \end{cases} \tag{3.2}$$

$$\frac{\partial u}{\partial t} - \sum_{i=1}^{n} \frac{\partial}{\partial x_i}\left(\frac{\partial u}{\partial x_i}\right)^{2k-1} = f \quad \text{dans } R^n \times \left]0, T\right[, \tag{3.3}$$

$$u(x,0) = 0. \tag{3.4}$$

Remarque 3.1

Comme à la Remarque 2.1, il résulte des hypothèses faites que

$$\frac{\partial u}{\partial t} \in L^{p'}(0, T; W^{-1,p'}(R^n))$$

de sorte que (3.4) a un sens.

3.2. Démonstration de l'unicité.

Soient u et w deux solutions du problème. Alors $w = u-v$ satisfait aux conditions analogues à (3.2), (3.4) et à

$$\frac{\partial w}{\partial t} - \sum_{i=1}^{n} \frac{\partial}{\partial x_i}\left(\psi\left(\frac{\partial u}{\partial x_i}, \frac{\partial v}{\partial x_i}\right) \frac{\partial w}{\partial x_i}\right) = 0 \tag{3.5}$$

où $$\psi(a,b) = \frac{a^{2k-1} - b^{2k-1}}{a-b}$$

Comme $w(t) \in L^p(W^{1,p}(R^n))$ et $w'(t) = \frac{\partial w}{\partial t} \in L^{p'}(W^{-1,p'}(R^n))$,
l'expression $(w'(t), w(t))$ a un sens p.p. (dualité entre $W^{1,p}(R^n)$ et $W^{-1,p'}(R^n)$) et définit une fonction sommable. De même

$$-\left(\frac{\partial}{\partial x_i}\psi\left(\frac{\partial u(t)}{\partial x_i}, \frac{\partial v(t)}{\partial x_i}\right)\frac{\partial w}{\partial x_i}, w(t)\right)$$ a-t-il un sens p.p. en t;

d'ailleurs, cette quantité vaut

$$\int_{R^n} \psi\left(\frac{\partial u}{\partial x_i}, \frac{\partial v}{\partial x_i}\right)\left(\frac{\partial w}{\partial x_i}\right)^2 dx ,$$

quantité $\geqslant 0$. Donc

$$(w'(t), w(t)) \leqslant 0$$

i.e. $|w(t)|^2$ décroissante. Comme $w(0) = 0$, on en déduit que $w \equiv 0$. L'unicité est démontrée.

3.3. Principe de la démonstration de l'existence.

On considère une suite f_m de $\mathcal{D}$ (dans l'ouvert $t > 0$) telle que

(3.6) $$f_m \longrightarrow f \text{ dans } L^{p'}(0,T;W^{-1,p'}(R^n)) .$$

Alors on sait (N.2) qu'il existe u_m satisfaisant aux conditions analogues à (3.2), (3.3), (3.4), avec f_m au lieu de f.

M. Visik montre (Visik, loc. cit., p. 315 et suivantes) que, lorsque $m \longrightarrow \infty$, les solutions u_m convergent vers la solution du problème.

4. Cas des problèmes mixtes.

Si l'on considère maintenant le problème mixte :

(4.1) $$\frac{\partial u}{\partial t} - \sum_{i=1}^{n} \frac{\partial}{\partial x_i}\left(\frac{\partial u}{\partial x_i}\right)^{2k-1} = f, \qquad x \in \Omega,\ t > 0,$$

où Ω est un ouvert borné de R^n, de frontière assez réguilère, avec

(4.2) $$u(x,0) = u_0(x) \quad \text{donné,}$$

et

(4.3) $$u(x,t) = 0 \qquad \text{si} \qquad x \in \Gamma,\ t > 0 \qquad (\Gamma = \text{frontière de } \Omega),$$

on a des résultats analogues. La méthode aussi est analogue, avec la différence suivante : sur un ouvert Ω, si l'on prend encore l'opérateur "séparant" B sous la forme (2.1), il apparaît, dans les intégrations par parties en x, des intégrales de surface; pour supprimer ces integrales, on prendra B sous la forme:

(4.4) $$Bu = -\psi \Delta u + u - \frac{\partial}{\partial t}(T - t)\frac{\partial u}{\partial t}$$

où ψ est une fonction régulière dans $\overline{\Omega}$, telle que

$$\begin{cases} \psi(x) > 0 & \text{si} \quad x \in \Omega, \\ \psi(x) = 0 & \text{si} \quad x \in \Gamma, \\ \dfrac{\partial \psi}{\partial n}(x) > 0 \ , & x \in \Gamma . \end{cases}$$

On considère alors, au lieu de (2.3); le système

$$(B^* \Lambda u_m(t), w_j) = (B^* f, w_j) \quad , \quad j = 1, \ldots\ldots, m$$

Nous renvoyons au travail de Visik pour les détails.

Commentaires sur le Chap. III.

Tous les résultats de ce Chapitre sont dûs à I. M. Visik, Mat. Sbornik, t. 59 (101), 1962, p. 289-325.

On trouvera dans cet article des résultats plus généraux, ce chapitre ayant pur seul but d'introduire à ce travail.

Le N. 1 montre que le méthodes "usuelles" - méthode de Faedo - Green - Galerkin, utilisées dans les problèmes non linéaires par E. Hopf, Math. Nachr. 4 (1951), p. 213-231 - ne conduisent pas ici à un résultat positif. Les méthodes utilisées dans les équations de Navier-Stokes et équations similaires (cf. en particulier J. L. Lions, C. R. Acad. Sc., t. 252 (1961), p. 657-659) donnent des estimations à priori sur $\frac{\partial u}{\partial t}$. Mais on a ici (dans l'exemple que nous choisissons) besoin d'estimations à priori sur

$$\frac{\partial^2 u}{\partial x_i \partial t} , \frac{\partial^2 u}{\partial x_i \partial x_j} .$$

C'est l'objet du N. 2, qui contient les idées essentielles introduites par I. M. Visik.

Le N. 3 donne un théorème d'existence et d'unicité et le N. 4 indique briévement comment étendre la méthode aux problèmes mixtes.

Chapitre IV

ÉQUATIONS LINÉAIRES DU DEUXIÈME ORDRE

1. Position du Problème.

1.1. On considère V et H comme au Chapitre I (et au Chapitre II, N. 4); V et H sont séparables.

On donne une famille de formes $a(t;u,v)$, $t \in (0,T)$, sesquilinéaires continues sur V ; on suppose que, pour tout $u, v \in V$, la fonction $t \longrightarrow a(t;u,v)$ est mesurable et que

(1.1) $$\left| a(t;u,v) \right| \leq M \|u\| \cdot \|v\| \qquad t \in (0,T).$$

On donne également une famille d'opérateurs $B(t) \in \mathcal{L}(H,H)$, tels que

(1.2) pour tout $f, g \in H$, $t \longrightarrow (B(t)f,g)$ est une fois continûment différentiable dans $[0,T]$.

On désignera par $\mathcal{W}$ l'espace des (classes de) fonctions u telles que

(1.3) $$u \in L^2(0,T;V)$$

(1.4) $$\frac{du}{dt} \in L^2(0,T;H)$$

(Pour le sens de (1.4), cf. Chap. I, N. 5).

Muni de la norme

$$\|u\|_{\mathcal{W}} = \left\{ \int_0^T (\|u(t)\|^2 + |u'(t)|^2)\, dt \right\}^{1/2}$$

$\mathcal{W}$ est un espace de Hilbert.

Notation : pour $u, v \in \mathcal{W}$, on pose:

(1.5) $$E(u,v) = \int_0^T \left[a(t;u(t),v(t)) - (u'(t),v'(t)) + ((B(t)u(t))', v(t)) \right] dt.$$

Comme on le vérifie sans peine, $E(u,v)$ est une forme sesquilinéaire continue sur $\mathcal{W}$.

Naturellement, si $u \in \mathcal{W}$ alors en particulier u est (p.p. égale à une fonction) continue de $[0,T] \longrightarrow H$. On pourra donc parler de $u(0)$, $u(T)$.

1.2 Le Problème.

On cherche $u \in \mathcal{W}$, satisfaisant à

(1.6) $u(0) = u_0$, u_0 donné dans V,

(1.7) $$E(u,\varphi) = \int_0^T (f(t), \varphi(t))\, dt + (u_1, \varphi(0)) \quad \text{pour tout } \varphi \in \mathcal{W}$$

tel que $\varphi(T) = 0$ où dans le deuxième membre de (1.7), f et u_1 sont donnés avec :

(1.8) $f \in L^2(0,T;H)$,

(1.9) $u_1 \in H$.

Naturellement, sans hypothèses supplémentaires, notamment sur $a(t;u,v)$, le problème précédent n'admet pas de solution.

Nous allons dans la suite donner des conditions suffisantes permettant d'affirmer l'existence et l'unicité d'une solution du problème précédent.

1. 3. Interprétation formelle du problème 1. 2.

Utilisant les opérateurs $A(t) \in \mathcal{L}(V;V')$ (cf. Chap. I et II) et intégrant formellement par parties dans (1.5), il vient :

(1.10) $A(t)\, u(t) + u''(t) + (B(t)\, u(t))' = f(t)$;

les conditions initiales sont, d'abord (1, 6) :

(1.6) $u(0) = u_o$

puis

(1.7) $u'(0) = u_1$

Nous allons justifier cela au N. suivant.

2. Propriétés des solutions (éventuelles) de (1.7).

Théorème 2. 1 Si $u \in \mathcal{W}^{\ell}$ est une solution du problème 1.2, alors elle a les propriétés suivantes :

(2.1) $u'' \in L^2(0, T;V')$

(2.2) $u'(0) = u_1$.

(Noter que (2. 1), joint au fait que $u' \in L^2(0, T;H)$, implique que n'est continue dans $[0, T] \longrightarrow V'$, de sorte que (2, 2) a un sens).

Démonstration.

On peut prendre dans (1, 7)

(2, 3) $\varphi(t) = \Psi(t)v$ $\Psi \in \mathcal{D}(]0, T[)$.

Alors (1.7) se réduit à

$$\int_o^T \left[a(t;u(t), v) + ((B(t)u(t))' , v)\right] \overline{\varphi(t)}dt - \int_o^T (u'(t), v)\ \overline{\varphi'(t)}dt =$$

$$= \int_o^T (f(t), v)\ \overline{\varphi(t)}dt$$

d'où, <u>au sens des distributions sur</u> $]0,T[$:

$$a(t;u(t),v) + ((B(t)u(t))', v) + \frac{d^2}{dt^2}(u(t), v) = (f(t), v), \quad \text{pour tout} \quad v \in V,$$

<u>ou encore</u>

$$A(t)u(t) + (B(t)u(t))' + u''(t) = f(t) \tag{2.4}$$

(au sens des distributions sur $]0,T[\longrightarrow V'$).

Mais on sait (cf. Chap. II) que $A(t)\, u \in L^2(0,T;V')$;par ailleurs $(B(t)\,u(t))' \in L^2(0,T;H)$ de sorte que (2, 4) implique (2, 1).

Mais alors, (cf. Chap. I, N.5, 5.3), si $\varphi \in \mathcal{W}$,

$$\int_0^T (u''(t), \varphi(t))\, dt = (u'(T), \varphi(T)) - (u'(0), \varphi(0)) - \int_0^T (u'(t), \varphi'(t))\, dt ;$$

si donc l'on prend le produit scalaire de (2, 4) avec $\varphi(t)$, $\varphi \in \mathcal{W}$, et $\varphi(T) = 0$, alors :

$$\int_0^T \Big[a(t;u(t), \varphi(t)) + ((B(t)\,u(t))', \varphi(t)) - (u'(t), \varphi'(t)) \Big]\, dt - (u'(0), \varphi(0)) =$$

$$= \int_0^T (f(t), \varphi(t))\, dt$$

d'où

$$E(u, \varphi) = \int_0^T (f(t), \varphi(t))\, dt + (u'(0), \varphi(0))$$

ce qui, en comparant avec (1.7) donne $u'(0) = u_1$ et achève la démonstration du théorème.

3. Théorème d'existence.

3.1. Théorème 3.1. On fait les hypothèses suivantes :

(3.1) $\begin{cases} t \longrightarrow a(t;u,v) \text{ est une fois continûment différentiable dans } [0,T], \\ u, v \in V ; a(t;u,v) = \overline{a(t;v,u)} \text{ et il existe } \lambda \text{ et } \alpha \geq 0 \text{ tels que} \\ a(t;v,v) + \lambda |v|^2 \geq \alpha \|v\|^2 , \quad v \in V ; \end{cases}$

(3.2) $\begin{cases} B(t) \text{ est hermitien dans } H \text{, pour tout } t \text{, et } t \longrightarrow (B(t)f,g) \text{ est} \\ \text{une fois continûment différentiable dans } [0,T] \text{ pour tout } f,g \in H. \end{cases}$

Alors, il existe une solution u et une seule du Problème 1.2 ayant en autre les propriétés suivantes :

(3.3) $u \in L^\infty(0,T;V)$

(3,4) $u' \in L^\infty(0,T;H)$

(et, naturellement, les propriétés données au Théorème 3.1).

3.2. Démonstration de l'existence.

Notens qu'un changement de u en $e^{kt}u$ change $a(t;u,v)$ en

$a(t;u,v) + k(B(t)u,v) + k^2(u,v)$ et $B(t)$ en $B(t) + 2kI$

(I = identité dans H). On peut donc toujours se ramener au cas où dans (3.1) $\lambda = 0$.

Soit $w_1 \ldots\ldots w_m \ldots$ une base de V . On définit $u_m(t)$, solution approchée d'ordre m , par

(3.5) $$u_m(t) = \sum_{i=1}^{m} g_{im}(t) w ;$$

où les $g_{im}(t)$ sont définis par le système différential (linéaire)

$$(3.6) \qquad \frac{d^2}{dt^2}(u_m(t), w_j) + (\frac{d}{dt}(B(t)u_m(t), w_j) + a(t;u_m(t), w_j) =$$

$$= (f(t), w_j) \qquad j = 1, 2, \ldots\ldots, m$$

avec les conditions initiales

$$(3.7) \qquad g_{im}(0) = \alpha_{im} \qquad g'_{im}(0) = \beta_{im}$$

où les α_{im} β_{im} sont choisis de façon que

$$(3.8) \qquad \begin{cases} \sum_i^m \alpha_{im} w_i \longrightarrow u_o & \text{dans } V \text{ lorsque } m \longrightarrow \infty, \\ \sum_i^m \beta_{im} w_i \longrightarrow u_1 & \text{dans } H \text{ lorsque } m \longrightarrow \infty. \end{cases}$$

Etablissons maintenant des majorations à priori pour $u_m(t)$.

Nous posons :

$$(3.9) \qquad \varphi_m(t) = u'_m(t)|^2 + a(t;u_m(t), u_m(t)),$$

$$(3.10) \qquad \psi_m(t) = |u'_m(t)|^2 + \|u_m(t)\|^2 .$$

Multiplions (3.6) par $\overline{g'_{jm}(t)}$ et sommons en j ;nous obtenons :

$$(u''_m(t), u'_m(t)) + a(t;u_m(t), u'_m(t)) + ((B(t)u_m(t))', u'_m(t)) =$$

$$= (f(t), u'_m(t)).$$

Prenons la relation complexe conjuguée et ajoutons ; si nous posons :

$$(3.11) \qquad \frac{d}{dt} a(t;u, v) = a'(t;u, v) \quad ,$$

nous obtenons :

$$\frac{d}{dt}\left[|u'_m(t)|^2 + a(t;u_m(t),u_m(t)) \right] + 2\mathrm{Re}((B(t)u_m(t))',u'_m(t)) -$$

$$- a'(t;u_m(t),u_m(t)) = 2\mathrm{Re}\,(f(t),u'_m(t)).$$

Utilisant (3.9) et intégrant, il vient :

$$\varphi_m(t) = \varphi_m(0) + \int_0^t \left[a'(\sigma,u_m(\sigma)\,u_m(\sigma)) - 2\,\mathrm{Re}((B(\sigma)u_m(\sigma))',u'_m(\sigma)) \right] d\sigma$$

$$+ 2\mathrm{Re} \int_0^t (f(\sigma),u'_m(\sigma))\,d\sigma$$

$$\leqslant \varphi_m(0) + c_1 \int_0^t (\psi_m(\sigma) + |f(\sigma)|^2)\,d\sigma$$

(les c_j désignent des constantes).

De l'expression de $\varphi_m(0)$ et de (3.8) on déduit

$$\varphi_m(0) \leqslant c_2(\|u_o\|^2 + |u_1|^2)$$

et comme $\varphi_m(t) \geqslant c_4\,\psi_m(t)$, il vient :

$$\text{(3.12)}\qquad \psi_m(t) \leqslant c_5(\|u_o\|^2 + |u_1|^2 + \int_0^t |f(\sigma)|\,d\sigma) + c_5 \int_0^t \psi_m(\sigma)\,d\sigma \leqslant$$

$$\leqslant c_5(\|u_o\|^2 + |u_1|^2 + \int_0^T |(f(\sigma)|^2\,d\sigma) + c_5 \int_0^t \psi_m(\sigma)\,d\sigma.$$

On en déduit, par un lemme classique de Gronwall :

$$\Psi_m(t) \leqslant c_5(\|u_o\|^2 + |u_1|^2 + \int_o^T |f(\sigma)|^2 d\sigma) \exp (c_5 t). \tag{3.13}$$

Conséquence :

u_m (resp. u'_m) demeure dans un ensemble borné de $L^\infty(0, T;V)$ (resp. de $L^\infty(0, T;H)$).

On peut donc extraire une suite $u_{m_\nu} = u_\nu$ telle que

$$\begin{cases} u_\nu \longrightarrow w \text{ dans } L^\infty(0, T;V) \text{ faible, (topologie faible de dual de } L^1(0, T;V)) \\ u'_\nu \longrightarrow w' \text{ dans } L^\infty(0, T;H) \text{ faible.} \end{cases} \tag{3.14}$$

Montrons que w est solution du problème. Pour cela, soit $\theta_j(t)$ une fonction une fois continûment différentiable dans $[0, T]$, avec $\theta_j(T) = 0$; utilisons (3.6) avec $m = \nu$ et j fixé $< \nu$; multiplions par $\overline{\theta_j(t)}$ et intégrons, posant

$$\chi(t) = \theta_j(t)\, w_j , \tag{3.15}$$

il vient, après intégration par parties :

$$E(u_\nu, \chi) = \int_o^T ((f(t), \chi(t))\, dt + (u'_\nu(o), \chi(o)). \tag{3.16}$$

Mais grâce à (3.14), $E(u_\nu, \chi) \longrightarrow E(w, \chi)$ et grâce à (3.8), 2ème relation, $(u'_\nu(o), \chi(o)) \longrightarrow (u_1, \chi(o))$. Donc :

$$E(w, \chi) = \int_o^T (f(t), \chi(t))\, dt + (u_1, \chi(o)) \tag{3.17}$$

et ceci pour toute χ de la forme (3.15) - donc aussi pour toute combinaison linéaire finie de telles fonctions et par passage à la limite, pour toute $\chi \in \mathcal{W}$ telle que $\chi(T) = 0$.

Par ailleurs de (3.14) suit que $u_\nu(o) \longrightarrow w(o)$ dans H faible (par ex) et comme d'après (3.8), 1ère relation, $u_\nu(o) \longrightarrow u_o$, on voit que $w(o) = u_o$ ce qui montre que w est solution du problème 1.2 et achève donc la démonstration de l'existence d'une solution ayant les propriétés (3.3), (3.4).

3.3. Démonstration de l'unicité.

Nous allons démontrer que si $u \in \mathcal{W}$, satisfait $u(o) = o$ et

(3.18) $\quad E(u, \varphi) = o$ pour tout $\varphi \in \mathcal{W}$, tel que $\varphi(T) = o$,

alors $u = o$.

Pour cela, nous prenons s avec $o < s < T$ et nous définissons $\varphi(t)$ par

$$\varphi(t) = \begin{cases} -\int_t^s u(\sigma)\,d\sigma & , \text{ si } t < s, \\ 0 & , \text{ si } t > s. \end{cases} \tag{3.19}$$

On a bien $\varphi \in L^2(0,T;V)$ et comme $\varphi' = u$, $\varphi'' \in L^2(0,T;V)$ donc $\in L^2(0,T;H)$.

Enfin $\varphi(T) = 0$. Donc on peut prendre cette φ dans (3.18); il vient:

$$\int_0^s \left[a(t;\varphi',\varphi) - (u',u) \right] dt + \int_0^s ((B(t)u(t))', \varphi(t))\, dt = 0. \tag{3.20}$$

La dernière intégrale vaut $-\int_0^s (B(t)u(t), u(t))\, dt$;prenant deux fois la partie réelle de (3.20), il vient :

$$\int_0^s \frac{d}{dt}(a(t;\varphi,\varphi) - |u(t)|^2)\, dt - \int_0^s \left[a'(t;\varphi,\varphi) + 2\mathrm{Re}(B(t)u(t),u(t))\right] dt = 0$$

d'où

$$a(0;\varphi(0),\varphi(0)) + |u(s)|^2 + \int_0^s \left[a'(t;\varphi,\varphi) + 2\mathrm{Re}(B(t)u(t),u(t))\right] dt = 0.$$

Il en resulte que

$$\|\varphi(0)\|^2 + |u(s)|^2 \leqslant c_6 \int_0^s (\|\varphi(t)\|^2 + |u(t)|^2)\, dt. \tag{3.21}$$

Mais si l'on pose

$$v(t) = \int_0^t u(\sigma)\, d\sigma, \tag{3.22}$$

alors $\varphi(t) = v(t) - v(s)$, $t < s$ et (3.21) s'écrit :

$$\|v(s)\|^2 + |u(s)|^2 \leqslant c_6 \int_0^s (\|v(t) - v(s)\|^2 + |u(t)|^2)\, dt \leqslant$$

$$2c_6 \int_0^s \|v(t)\|^2\, dt + 2\, c_6\, s\, \|v(s)\|^2 + c_6 \int_0^s |u(t)|^2\, dt.$$

Soit s_0 tel que, par exemple, $1 - 2\, c_6\, s_0 = 1/2$. Alors, pour $0 \leqslant s \leqslant s_0$, on a :

$$\|v(s)\|^2 + |u(s)|^2 \leqslant c_7 \int_0^s (\|v(t)\|^2 + |u(t)|^2)\, dt$$

donc $u = 0$ dans $(0, s_0)$. En recommencant, on trouve que $u = 0$ dans $(s_0, 2s_0)$ etc.

Ceci achève la démonstration du théorème 3.1.

4. Propriétés supplémentaires de la solution.

4.1. Théorème 4.1. Hypothèses du Théorème 3.1. Soit u la solution du Problème 1.2. D'après les Théorèmes 2.1 et 3.1 nous savons que u est p.p. égale à une fonction, notée u_*, continue de $[0, T] \to H$ (et même $\to V^{1/2} H^{1/2}$, cf. commentaires), de derivée u'_* continue de $[0, T] \to V'$ (et même $\to H^{1/2} (V')^{1/2}$).

En autre :

(4.1) $\qquad u_*(t) \in V \quad$ pour tout $\quad t \in [0, T]$

et

(4.2) $\qquad t \to ((u_*(t), v))$ est continue dans $[0, T]$ pour tout $v \in V$;

(4.3) $\qquad \frac{d}{dt} u_*(t) \in H \quad$ pour tout $\quad t \in [0, T]$

et

(4.4) $\qquad t \to (\frac{du_*(t)}{dt}, g)$ est continue dans $[0, T]$ pour tout $g \in H$;

enfin

(4.5)
$$\begin{cases} \text{quels que soient } v \in V,\ g \in H, \text{ l'application} \\ \{u_0, u_1, f\} \to \{((u_*(t), v)), (\frac{d}{dt} u_*(t), g)\} \\ \text{est continue de } V \times H \times L^2(0, T; H) \to C(0, T) \times C(0, T) \end{cases}$$

(où $C(0, T)$ = fonctions continues dans $[0, T]$, topologie de la convergence uniforme).

4.2. - Démonstration de (4.1) et (4.3).

Pour $u \in \mathcal{W}$, $v \in V$, nous poserons :

(4.6) $$\mathcal{M}(u(t),v) = a(t;u(t),v) + ((B(t)u(t))',v) + \frac{d^2}{dt^2}(u(t),v)$$

ce qui est une distribution sur $]0,T[$. En prenant toutes les dérivées au sens des distributions, (4.6) a un sens pour $u \in L^2(0,T;V)$.

On a, si u est solution du Problème 1.2:

(4.7) $$\mathcal{M}(u(t),v) = (f(t),v) \qquad \text{pour tout } v \in V, \text{ ou}$$

(4.7bis) $$\mathcal{M}(u_*(t),v) = (f(t),v).$$

Soit maintenant τ un nombre fixé , $0 < \tau < T$.

Designons par :

Y_τ la fonction nulle pour $t < \tau$, = 1 pour $t > \tau$;

$\delta_{(\tau)}$ = masse + 1 au point τ ;

$\delta'_{(\tau)} = \frac{d}{dt}\delta_{(\tau)}$

Alors

(4.8) $$\begin{cases} \mathcal{M}(Y_\tau u_*(t),v) = (Y_\tau f(t),v) + (u_*(\tau),v)\,\delta'_{(\tau)} + \\ \qquad + \left[(B(\tau)u_*(\tau),v) + (u'_*(\tau),v)\right]\delta_{(\tau)} \end{cases}$$

D'un autre coté, revenant à la démonstration du Théorème 3.1, on voit, en utilisant (3.13), que l'on peut supposer que la suite u_ν ayant les propriétés (3.14) vérifie en autre

(4.9) $$\begin{cases} u_\nu(\tau) \longrightarrow \chi_0 \text{ dans } V \text{ faible} \\ u'_\nu(\tau) \longrightarrow \chi_1 \text{ dans } H \text{ faible} \end{cases}$$

et

(4.10) $$\|\chi_0\|^2 + |\chi_1|^2 \leq c_5(\|u_0\|^2 + |u_1|^2 + \int_0^T |f(\sigma)|^2 d\sigma) \exp(c_5 T).$$

Mais

$$\mathcal{M}(Y_\tau u_\nu(t), w_j) = (Y_\tau f(t), w_j) + (u_\nu(\tau), w_j)\delta'_{(\tau)} + $$
$$+ \left[(B(\tau)u_\nu(\tau), w_j) + (u'_\nu(\tau), w_j)\right]\delta_{(\tau)}$$

et en passant à la limite selon ν, on obtient (en utilisant (4.9)):

$$\mathcal{M}(Y_\tau u(t), w_j) = (Y_\tau f(t), w_j) + (\chi_0, w_j)\delta'_{(\tau)} +$$
$$+ \left[(B(\tau)\chi_0, w_j) + (\chi_1, w_j)\right]\delta_{(\tau)}$$

Cette realtion a alors lieu pour toute combinaison linéaire finie des w_j et donc

(4.11) $$\begin{cases} \mathcal{M}(Y_\tau u(t), v) = (Y_\tau f(t), v) + (\chi_0, v)\delta'_{(\tau)} + \\ \qquad + \left[(B(\tau)\chi_0, v) + (\chi_1, v)\right]\delta_{(\tau)} \text{ pour tout } v \in V. \end{cases}$$

Evidemment $\mathcal{M}(Y_\tau u(t), v) = \mathcal{M}(Y_\tau u_*(t), v)$ et donc (4.8) (4.11) impliquent

$$u_*(\tau) = \chi_0, \qquad u'_*(\tau) = \chi_1$$

et comme $\chi_0 \in V$, $\chi_1 \in H$, on a (4.1) et (4.2). En outre, (4.10) donne :

(4.12) $$\|u_*(t)\|^2 + |u'_*(t)|^2 \leq c_5(\|u_0\|^2 + |u_1|^2 + \int_0^T |f(\sigma)|^2 d\sigma) \text{esp}(c_5 T).$$

4.3. <u>Démonstration de (4.2) (4.4).</u>

Soit τ_m une suite de (0, T), telle que $\tau_m \to \beta$, $0 < \beta < T$.

D'après (4.12), on voit que l'on peut estraire une suite σ_n de τ_n telle que

(4.13) $$\begin{cases} u_*(\sigma_n) \longrightarrow \xi_0 \text{ dans } V \text{ faible,} \\ u'_*(\sigma_n) \longrightarrow \xi_1 \text{ dans } H \text{ faible.} \end{cases}$$

Nous allons démontrer que

(4.14) $$\xi_0 = u_*(\beta), \qquad \xi_1 = u'_*(\beta).$$

Alors (4.2) et (4.4) en résultent.

Pour montrer (4.14), notons d'abord que $Y_{\sigma_n} u \longrightarrow Y_\beta u$ dans $L^2(0,T;V)$; donc $a(t;Y_{\sigma_n} u(t),v) \longrightarrow a(t;Y_\beta u(t),v)$ dans $L^2(0,T)$. Puis $B(t)Y_{\sigma_n} u \longrightarrow B(t)Y_\beta u$ dans $L^2(0,T;H)$ donc $((B(t)Y_{\sigma_n} u)',v) \longrightarrow ((B(t)Y_\beta u)',v)$ au sens des distributions sur $]0,T[$. De même, $\frac{d^2}{dt^2}(Y_{\sigma_n} u(t),v) \longrightarrow \frac{d^2}{dt}(Y_\beta (u(t),v)$ dans $\mathcal{D}'(0,T)$.

Donc

(4.15) $$\mathcal{M}(Y_{\sigma_n} u(t),v) \longrightarrow \mathcal{M}(Y_\beta u(t),v) \qquad \text{dans } \mathcal{D}'(0,T).$$

Mais

$$\mathcal{M}(Y_{\sigma_n} u(t),v) = (Y_{\sigma_n} f(t),v) + (u_*(\sigma_n),v)\,\delta'_{(\sigma_n)} + \\ + \left[(B(\sigma_n)u_*(\sigma_n),v) + (u'_*(\sigma_n),v)\right]\delta_{(\sigma_n)}$$

et d'après (4.13), ceci converge vers

$$(Y_\beta f(t),v) + (\xi_0,v)\delta'_{(\beta)} + (B(\beta)\xi_0 + \xi_1,v)\,\delta_{(\beta)}.$$

Utilisant (4.15), on en tire

(4.16) $$\mathcal{M}(Y_\beta u(t),v) = (Y_\beta f(t),v) + (\xi_0,v)\delta'_{(\beta)} + (B(\beta)\xi_0 + \xi_1,v)\,\delta_{(\beta)}.$$

Comparant avec (4.8) (pour $\tau = \beta$), il vient (4.14), d'où (4.2) (4.4).

La démonstration précédente suppose $\beta \neq 0$. Pour $\beta = 0$, il suffit de faire la modofication suivante : on désigne par $\tilde{u}$ le prolongement de u par 0 pour $t < 0$ et on désigne par $\tilde{a}(t;u,v)$, $\tilde{B}(t)$ le prolongement de $a(t;u,v)$, $B(t)$ également par 0 pour $t < 0$. Si l'on pose :

$$\mathcal{M}(\tilde{u}(t),v) = \tilde{a}(t;\tilde{u}(t),v) + ((\tilde{B}(t)\,\tilde{u}(t))',v) + \frac{d^2}{dt^2}(\tilde{u}(t),v)$$

alors

$$\mathcal{M}(\tilde{u}(t),v) = (\tilde{f}(t),v) + (u_*(0),v)\,\delta'_{(0)} + (B(0)u_*(0) + u'_*(0),\ v)\,\delta_{(0)}\,.$$

On introduit σ_n comme précédemment et l'on vérifie que

$$\mathcal{M}(Y_{\sigma_n}\tilde{u}(t),v) \longrightarrow \mathcal{M}(\tilde{u}(t),v) \quad \text{dans } \mathcal{D}'(-\infty,T) = \text{espace des}$$

distributions sur $]-\infty,T[$. Mais, comme précédemment, $u_*(\sigma_n) \rightarrow \xi_0$ dans V faible, $u'_*(\sigma_n) \longrightarrow \xi_1$ dans H faible, et donc

$$\mathcal{M}(Y_{\sigma_n}\tilde{u}(t),v) \longrightarrow (\tilde{f}(t),v) + (\xi_0,v)\,\delta'_{(0)} + (B(0)\xi_0 + \xi_1, v)\,\delta_{(0)}$$

d'où $u_*(0) = \xi_0$, $u'_*(0) = \xi_1$, d'où encore le résultat désiré.

4.4. Démonstration de (4.5)

Il suffit de montrer que le graphe de l'application est fermé. Soit donc une suite $\left\{u_0^{(m)},\ u_1^{(m)},\ f^{(m)}\right\} \longrightarrow 0$ dans $V \times H \times L^2(0,T;H)$, telle que

$$\left\{((u_*^{(m)}(t),v)),\ (\frac{d}{dt}u_*^{(m)}(t),v)\right\} \quad \text{converge dans} \quad C(0,T) \times C(0,T).$$

Comme $u^{(m)} \longrightarrow 0$ dans $L^\infty(0,T;V)$, $\frac{du^{(m)}}{dt} \longrightarrow 0$ dans $L^\infty(0,T;H)$, il en résulte que la limite de $\left\{((u_{*}^{(m)}(t), v)), (\frac{d}{dt} u_{*}^{(m)}(t), v)\right\}$ est nécessairement $\left\{0, 0\right\}$, d'où (4.5), ce qui achève la démonstration du théorème.

4.5. <u>Remarque</u>. De (4.12) résulte, pour tout t fixé, l'application $\left\{u_0, u_1, f\right\} \longrightarrow \left\{u_{*}(t), u_{*}'(t)\right\}$ est continue de $V \times H \times L^2(0, T; V)$ dans $V \times H$ fort.

5. <u>Propriété de dépendance continue de la solution en</u> $a(t;u,v)$ <u>et</u> $B(t)$.

5.1. On se donne une <u>famille</u> de formes $a^{\beta}(t;u,v)$ et d'operateurs $B^{\beta}(t)$, $\beta = 1, 2, \ldots,$ ayant les propriétés suivantes :

(5.1) pour chaque β, $a^{\beta}(t;u,v)$ est une fois continûment différentiable dans $[0, T]$, pour tout $u, v \in V$, et il existe λ et $\alpha > 0$ tels que : $a^{\beta}(t;v,v) + \lambda |v|^2 \geq \alpha \|v\|^2$, $v \in V$ (λ et α indépendants de β) ;

(5.2) $|a^{\beta}(t;u,v)| + |\frac{d}{dt} a^{\beta}(t;u,v)| \leq M \|u\| \cdot \|v\|$, M étant une constante indépendente de t et de β ;

(5.3) $a^{\beta}(t;u,v) \longrightarrow a(t;u,v)$, pour tout $t \in [0, T]$ fixé, $\beta \to \infty$, $u \in V$ fixé, uniformément pour $v \in$ borné de V ;

(5.4) $B^{\beta}(t) \in \mathcal{L}(H;H)$, hermitien; $t \longrightarrow (B^{\beta}(t)f, g)$ est une fois continûment différentiable dans $[0, T]$ pour tout $f, g \in H$;

(5.5) $|B^{\beta}(t)| + |\frac{d}{dt} B^{\beta}(t)| \leq M$; $B^{\beta}(t)f \longrightarrow B(t)f$ dans H fort, t fixé, $\beta \to \infty$, dans $[0, T]$, f fixé dans H ; au même sens, $\frac{d}{dt} B^{\beta}(t) f \longrightarrow \frac{d}{dt} B(t) f$.

On a le résultat suivant ; en désignant par u^β la solution de

(5.6) $$E^\beta(u^\beta, \varphi) = \int_0^T (f(t), \varphi(t))\, dt + (u_1, \varphi(0)) ,$$

où

(5.7) $$E^\beta(u, \varphi) = \int_0^T \left[a^\beta(t;u(t), \varphi(t)) - (u'(t), \varphi'(t)) + ((B\ (t)u(t))', \varphi(t))\right] dt,$$

$u^\beta, \varphi \in \mathcal{W}$, $\varphi(T) = 0$, et avec

(5.8) $$u^\beta(0) = u_0 .$$

[N.B. On pourrait aussi prendre une famille $f^\beta, u_0^\beta, u_1^\beta$].

Théorème 5.1. On suppose que (5.1).....(5.5) ont lieu. Alors lorsque $\beta \to \infty$, on a :

(5.9) $$u^\beta \to u \text{ dans } L^\infty(0, T;V) \text{ faible (dual faible de } L^1(0, T;V)),$$

(5.10) $$\frac{du^\beta}{dt} \to \frac{du}{dt} \text{ dans } L^\infty(0, T;H) \text{ faible,}$$

(5.11) $$\frac{d^2u^\beta}{dt^2} \to \frac{d^2u}{dt^2} \text{ dans } L^2(0, T;V') \text{ faible,}$$

(5.12) $$u_*^\beta(t) \to u_*(t) \text{ dans } V \text{ faible, lorsque } t \text{ est fixé,}$$

et

(5.13) $$(u_*^\beta(t), v) \to (u_*(t), v) \text{ uniformément en } t \text{, pour tout}$$

v fixé dans V,

(5.14) $$\frac{d}{dt}u_*^\beta(t) \to \frac{d}{dt}u_*(t) \text{ dans } H \text{ faible, lorsque } t \text{ est fixé,}$$

(5.15) $$\begin{cases} (B^\beta(t)u_*^\beta(t) + \frac{d}{dt}u_*^\beta(t), v) \longrightarrow (B(t)u_*(t) + \frac{d}{dt}u_*(t), v) \\ \text{pour } v \text{ fixé, uniformément en } t. \end{cases}$$

Démonstration.

1) D'après (4.12) et les hypothèses faites sur a^{β}, B^{β}, on a :

$$(5.16)\qquad \| u_*^{\beta}(t)\|^2 + \left|\frac{d}{dt} u_*^{\beta}(t)\right|^2 \leq C_1 \left(\| u_o\|^2 + |u_1|^2 + \int_o^T |f(\sigma)|^2 d\sigma\right) = C_2$$

(où les C_i désignent des constantes, différentes de celles du N. 4; les C_i sont en particulier indépendants de β).

Donc on peut extraire une suite β' de β telle que

$$u^{\beta'} \longrightarrow w \quad \text{dans } L^{\infty}(0,T;V) \text{ faible (dual faible de } L^1(0,T;V)),$$

$$\frac{d}{dt} u^{\beta'} \longrightarrow \frac{dw}{dt} \quad \text{dans } L^{\infty}(0,T;H) \text{ faible.}$$

La première chose à montrer est que $w = u$; alors (5.4) (5.10) sera démontré.

Comme $\int_o^T (u'^{\beta}(t), \varphi'(t))\,dt \rightarrow \int_o^T (w'(t), \varphi'(t))\,dt$, et comme $u^{\beta}(0) = u_o$, il nous reste à montrer que

$$(5.17)\qquad \int_o^T a^{\beta'}(t; u^{\beta'}(t), \varphi(t))\,dt \rightarrow \int_o^T a(t; w(t), \varphi(t))\,dt$$

et que

$$(5.18)\qquad \int_o^T ((B^{\beta'}(t)u^{\beta'}(t))', \varphi(t))\,dt \rightarrow \int_o^T ((B(t)w(t))', \varphi(t))\,dt.$$

Démonstration de (5.17)

On peut écrire : $a^{\beta}(t;u,v) = ((\mathcal{A}^{\beta}(t)u, v)) = ((u, \mathcal{A}^{\beta}(t)v))$ (car $a^{\beta}(t;u,v) =$

$= \overline{a^{\beta}(t;v,u))}$ et donc (5.17) est vrai si l'on a :

(5.19) $\mathcal{A}^{\beta}(t)\varphi(t) \longrightarrow \mathcal{A}(t)\,\varphi(t)$ dans $L^2(0,T;V)$ fort, $\varphi \in L^2(0,T;V)$.

Or $\|\mathcal{A}^{\beta}(t)\,\varphi(t)\| \leq M\,\|\varphi(t)\|$ et d'après (5.3), $\mathcal{A}^{\beta}(t)\,\varphi(t) \longrightarrow$ $\longrightarrow \mathcal{A}(t)\,\varphi(t)$ dans V fort, p.p. en t, d'où (5.19).

Démonstration de (5.18).

Par intégration par parties, l'expression $\int_0^T ((B^{\beta}(t)u^{\beta}(t))',\ \varphi(t))dt$ (on écrit β au lieu de β') vaut :

$$- (B^{\beta}(0)u_0,\ \varphi(0)) - \int_0^T (B^{\beta}(t)u^{\beta}(t),\ \varphi'(t))\,dt =$$

$$= - (B^{\beta}(0)u_0,\ \varphi(0)) - \int_0^T (u^{\beta}(t), B^{\beta}(t)\ \varphi'(t))\,dt$$

et ceci converge vers $- (B(0)u_0,\ \varphi(0)) - \int_0^T (w(t), B(t)\ \varphi'(t))\,dt$ (utiliser (5.5)).

D'où (5.18).

2) Démonstration de (5.11).

Chaque u^{β} satisfait à

$$A^{\beta}(t)u^{\beta}(t) + \frac{d}{dt}\,(B^{\beta}(t)u^{\beta}(t)) + \frac{d^2}{dt^2}\,u^{\beta}(t) = f(t) \quad \text{dans} \quad]0,T[\ .$$

On a donc (5.11) si l'on vérifie que

(5.20) $A^{\beta}(t)u^{\beta}(t) \longrightarrow A(t)u(t)$ dans $L^2(0,T;V')$ faible,

(5.21) $\qquad \frac{d}{dt}(B^\beta(t)u^\beta(t)) \longrightarrow \frac{d}{dt}(B(t)u(t))$ dans $L^2(0,T;V')$ faible.

Pour (5.20) on note que, si $\psi \in L^2(0,T;V)$, on a :

$$\int_0^T (A^\beta(t)u^\beta(t), \psi(t))\, dt = \int_0^T a^\beta(t; u^\beta(t), \psi(t))\, dt = \int_0^T ((u^\beta(t), \mathcal{A}^\beta(t)\psi(t)))\, dt$$

et le résultat suit comme dans la démonstration de (5.17).

Pour (5.21) , on note que

$$\frac{d}{dt}(B^\beta(t)u^\beta(t)) = (\frac{d}{dt}B^\beta(t))u^\beta(t) + B^\beta(t)\frac{d}{dt}u^\beta(t)$$

et on en déduit (5.24) en utilisant (5.5).

3) <u>Démonstration de</u> (5.12), (5.13), (5.14), (5.15).

On introduit maintenant (cf. N.4) :

(5.22) $\qquad \mathcal{M}^\beta(u(t),v) = a^\beta(t;u(t),v) + \frac{d}{dt}(B^\beta(t)u(t),v) + \frac{d^2}{dt^2}(u(t),v)$

ce qui a un sens pour $u \in L^2(0,T;V)$ et on note (cf. (4.8)) que :

(5.23) $\mathcal{M}^\beta(Y_\sigma u^\beta(t),v) = (Y_\sigma f(t),v) + (u_*^\beta(\sigma),v)\delta'_{(\sigma)} + (B^\beta(\sigma)u_*^\beta(\sigma) + u'^\beta_*(\sigma),v)\,\delta_{(\sigma)}$.

Supposons σ_0 <u>fixé</u> dans $]0,T[$. D'après (5.16), on peut extraire une suite de β , que nous désignerons encore par β pour simplifier, telle que

(5.24) $\qquad \begin{cases} u_*^\beta(\sigma_0) \longrightarrow \chi_0 & \text{dans } V \text{ faible,} \\ \frac{d}{dt}u_*^\beta(\sigma_0) \rightarrow \chi_1 & \text{dans } H \text{ faible} \end{cases}$

Admettons pour un instant le

Lemme 5.1 Lorsque $\beta \to \infty$, $\mathcal{M}^\beta(Y_\sigma u^\beta(t), v) \to \mathcal{M}(Y_\sigma u(t), v)$ dans $\mathcal{D}'(0,T)$ (distribution sur $]0,T[$), uniformément pour $\sigma \in [0,T]$.

D'après (5.23), il résulte du Lemme 5.1 que

(5.25) $$(Y_\sigma f(t), v) + (u_*^\beta(\sigma), v)\delta'_{(\sigma)} + (B^\beta(\sigma)u_*^\beta(\sigma) + u'^\beta_*(\sigma), v)\delta_{(\sigma)}$$

converge vers

(5.26) $$(Y_\sigma f(t), v) + (u_*(\sigma), v)\delta'_{(\sigma)} + (B(\sigma)u_*(\sigma) + u'_*(\sigma), v)\delta_{(\sigma)}$$

dans $\mathcal{D}'(0,T)$, uniformément pour $\sigma \in [0,T]$.

Donc :

$$(u_*^\beta(\sigma), v) \to (u_*(\sigma), v) \quad \text{uniformément en } \sigma,\ v \in V,$$

$$(B^\beta(\sigma)u_*^\beta(\sigma) + u'^\beta_*(\sigma), v) \to (B(\sigma)u_*(\sigma) + u'_*(\sigma), v) \text{ unifor-}$$

mément en σ, ce qui montre (5.13) (5.15). Par ailleurs, d'après (5.24) l'expression (5.25) converge (pour $\sigma = \sigma_0$) vers

$$(Y_{\sigma_0} f(t), v) + (\chi_0, v)\delta'_{(\sigma_0)} + (B(\sigma_0)\chi_0 + \chi_1, v)\delta_{(\sigma_0)}$$

d'où en comparant avec l'expression (5.26), $u_*(\sigma_0) = \chi_0$, $u'_*(\sigma_0) = \chi_1$.

Donc $u_*^\beta(\sigma_0) \to u_*(\sigma_0)$ dans V faible, $u'^\beta_*(\sigma_0) \to u'_*(\sigma_0)$ dans H faible, pour une suite extraite; mais la limite étant indépendante de la suite extraite, on a le résultat (5.12) (5.14).

Le théorème est donc complètement démontré sous réserve de la Vérification du Lemme 5.1.

Il y a trois choses à démontrer :

(5.27) $a^\beta(t; Y_\sigma u^\beta(t), v) \to a(t; Y_\sigma u(t), v)$ dans $\mathcal{D}'(0,T)$ faible, uniformément en σ ;

(5.28) $((B^{\beta}(t)Y_{\sigma}u^{\beta}(t))',v) \to ((B(t)Y_{\sigma}u(t))',v)$ au même sens que dans (5.27);

(5.29) $\frac{d^2}{dt^2}(Y_{\sigma}u^{\beta}(t),v) \to \frac{d^2}{dt^2}(Y_{\sigma}u(t),v)$ au même sens que dans (5.27).

Démonstration de (5.27).

Soit $\psi \in \mathcal{D}(0,T)$. Il faut montrer que

$$\int_0^T a\,(t;Y_{\sigma}u^{\beta}(t),\psi(t)v)dt \to \int_0^T a(t;Y_{\sigma}u(t),\psi(t)v)dt$$

uniformément en σ.

Le premier membre vaut $\int_0^T ((u^{\beta}(t),Y_{\sigma}\mathcal{A}^{\beta}(t)\psi(t)v))\,dt =$

$$= \int_0^T ((u^{\beta}(t),(\mathcal{A}^{\beta}(t)-\mathcal{A}(t))\,Y_{\sigma}\,\psi(t)v))\,dt +$$

$$+ \int_0^T ((u^{\beta}(t),\mathcal{A}(t)Y_{\sigma}\psi(t)v))\,dt.$$

La dernière expression converge vers $\int_0^T ((u(t),\mathcal{A}(t)Y_{\sigma}\,\psi(t)v))dt$ uniformément en σ pour que, lorsque σ varie, $\mathcal{A}(t)Y_{\sigma}\,\psi(t)v$ demeure dans un compact de $L^2(0,T;V)$.

On a donc le résultat si

$$\int_0^T ((u^{\beta}(t),(\mathcal{A}^{\beta}(t)-\mathcal{A}(t))\,Y_{\sigma}\,\psi(t)v))\,dt \to 0$$

uniformément en σ. Or

$$\int_0^T \| (\mathcal{A}^\beta(t) - \mathcal{A}(t)) Y_\sigma \psi(t) v \|^2 dt \leq \int_0^T \| (\mathcal{A}^\beta(t) - \mathcal{A}(t)) \psi(t) v \|^2 dt \to 0$$

d'où le résultat.

Commentaires sur le Chapitre IV.

Les résultats de ce Chapitre complètent quelque peu les résultats donnés dans Lions, Equations différentielles opérationelles, Springer, Collection Jaune, t. 111, 1961. Les résultats des N. 3 et 4 ont été donnés à l'Université de Montréal, Ecole d'Eté, Juillet 1962.

Il serait intéressant d'affaiblir les hypothèses de différentiabilité faites sur $a(t;u,v)$; $a(t;u,v)$ continue et $\frac{d}{dt} a(t;u,v)$ mesurable et bornée suffit (même démonstration que celle du texte !); mais il serait par ex. intéressant de savoir s'il est suffisant, pour l'existence et l'unicité, que $a(t;u,v)$ vérifie une condition de Lipschitz en t.

Nous n'avons que des résultats fragmentaires si nous remplaçons V par une famille d'espace $V(t)$ comme au Chap. II.

Le résultat de dépendance continue en les "coefficients" donné au N. 5 peut être complété par un résultat de dépendance continue en les espaces, mais cela n'est pas donné ici. Les résultats des N. 4 et 5 apportent des réponses (peut être insuffisantes....) à des questions posées par M. M. L. Amerio et S. Zaidman.

On peut donner une démonstration de l'existence d'une solution (cf. Th. 3. 1.) par un procédé de perturbation singulière en considérant le problème comme limite de problèmes "elliptiques" (dans un certain sens...) (nous avons utilisé un procédé de ce genre pour la démonstration du théorème d'existence dans le cas des équations différentielles opérationelles du 1er ordre). En voici le principe. On se place sur la demidroite $(0, +\infty)$; $a(t;u,v)$, $B(t)$ sont supposés donnés sur $(0, \infty)$ avec :

$$\left|a(t;u,v)\right| + \left|a'(t;u,v)\right| \leq M \|u\| \|v\| , \qquad t \geq 0,$$
$$|B(t)| + |B'(t)| \leq C_1, \qquad t \geq 0,$$
$$a(t;v,v) \geq \alpha \|v\|^2, \quad \alpha > 0, \quad t \geq 0.$$

Notons que, dans ces conditions :

$$(1) \qquad 2\,\mathrm{Re} \int_0^\infty ((B(t)\psi(t))', e^{-2\gamma t}\psi'(t))\,dt \leq C_2 \int_0^\infty (\|e^{-\gamma t}\psi\|^2 + |e^{-\gamma t}\psi'|^2)\,dt$$

(on peut même remplacer dans le $2^{\text{ème}}$ membre la norme $\| \ \|$ par la norme $| \ |$), où $\gamma > 0$ quelconque, C_2 indépendant de γ.

Alors on choisit γ de façon que

$$(2) \qquad \begin{cases} 2\gamma a(t;v,v) - a'(t;v,v) - C_2\|v\|^2 \geq \alpha_1\|v\|^2, \quad \alpha_1 > 0, \ v \in V, t \geq 0 \\ \gamma \geq C_2 + \delta, \quad \delta > 0. \end{cases}$$

On désigne par W_γ l'espace des u avec $(L_+^2(X) = L^2(0,\infty;X))$:

$$(3) \qquad e^{-\gamma t}u \in L_+^2(V), \ e^{-\gamma t}u' \in L_+^2(V), \ e^{-\gamma t}u'' \in L_+^2(H), u(0) = 0.$$

Pour $u, v \in W_\gamma$, on pose :

$$(4) \qquad \Pi_\varepsilon(u,v) = \int_0^\infty a(t;u(t), e^{-2\gamma t}v'(t))\,dt + \varepsilon\int_0^\infty ((e^{-\gamma t}u'(t), e^{-\gamma t}v'(t)))\,dt +$$
$$+ \int_0^\infty ((B(t)u(t))', e^{-2\gamma t}v'(t))\,dt - \int_0^\infty (u', (e^{-2\gamma t}v')')\,dt +$$
$$+ \varepsilon\int_0^\infty (u'', (e^{-2\gamma t}v')')\,dt, \quad \varepsilon > 0.$$

On vérifie que, pour $\varepsilon \leq 1/4\gamma$, on a :

$$(5)\quad \begin{cases} \operatorname{Re} \Pi_\varepsilon(v,v) \geq \beta \int_0^\infty (\|e^{-\gamma t} v\|^2 + |e^{-\gamma t} v'|^2)\,dt + \beta\,|v'(0)|^2 + \\ + \varepsilon \int_0^\infty (\|e^{-\gamma t} v'\|^2 + |e^{-\gamma t} v''|^2)\,dt, \qquad \beta > 0, \qquad v \in W_\gamma . \end{cases}$$

Alors, <u>il existe</u> $u_\varepsilon \in W_\gamma$ <u>unique, tel que</u>

$$(6)\qquad \Pi_\varepsilon(u_\varepsilon, v) = \int_0^\infty (e^{-\gamma t} f, e^{-\gamma t} v')\,dt + (u_1, v'(0)),\ \underline{\text{pour tout}}\ v \in W_\gamma ,$$

u_1 donné dans H, f donné avec $e^{-\gamma t} f \in L^2_+(H)$.

On montre ensuite que lorsque $\varepsilon \longrightarrow 0$, u_ε converge au sens suivant :

$$e^{-\gamma t} u_\varepsilon \longrightarrow e^{-\gamma t} u \quad \text{dans} \quad L^2_+(V) \ \text{faible},$$

$$e^{-\gamma t} u'_\varepsilon \longrightarrow e^{-\gamma t} u' \quad \text{dans} \quad L^2_+(H) \ \text{faible},$$

où u satisfait à

$$(7)\quad \begin{cases} \int_0^\infty \left[a(t; u(t), e^{-2\gamma t} \varphi') + ((B(t)u(t))', e^{-2\gamma t} \varphi'(t)) - (u', (e^{-2\gamma t} \varphi')') \right] dt = \\ \qquad = \int_0^\infty (f(t), e^{-2\gamma t} \varphi')\,dt + (u_1, \varphi'(0)) \end{cases}$$

pour tout $\varphi \in W_\gamma$ et

$$(8)\qquad u(0) = 0$$

On vérifie comme dans notre livre, Chap. VIII, p. 152, 153, <u>que la</u>

restriction de u à l'intervalle (0; T) est solution du problème 1.2 avec u(0) = 0.

On a donc obtenu l'existence d'une solution comme limite des problèmes " elliptiques" (6).

CENTRO INTERNAZIONALE MATEMATICO ESTIVO

(C. I. M. E.)

L. NIRENBERG

EQUAZIONI DIFFERENZIALI ORDINARIE NEGLI SPAZI DI BANACH

ROMA - Istituto Matematico dell'Università

EQUAZIONI DIFFERENZIALI ORDINARIE NEGLI SPAZI DI BANACH

di

L. NIRENBERG

Capitolo I

INTRODUZIONE

1.1. Ci proponiamo di descrivere i risultati di un recente lavoro in collaborazione con Agmon [2] . Questo lavoro riguarda lo studio delle equazioni della forma

$$L\,u = \frac{1}{i}\,\frac{du}{dt} - Au = f \tag{1.1}$$

e in particolare il comportamento delle soluzioni quando $t \to +\infty$. Le funzioni assumono i loro valori in uno spazio di Banach. Noi non tratteremo il problema dei valori iniziali: per la classe di equazioni considerata questo problema non è ben posto. Infatti noi tratteremo equazioni che provengono da equazioni differenziali a derivate parziali in un cilindro che ha l'asse t come generatrice, per esempio equazioni ellittiche. (L'operatore A rappresenta un operatore differenziale a derivate parziali nelle altre variabili). Quindi noi consideremo proprietà delle soluzioni, non l'esistenza di esse.

Parecchie delle questioni qui considerate sono state suggerite da ricerche dovute a Lax [8] , [9] , [10] .

In questo capitolo noi descriveremo i problemi e mostreremo come le condizioni richieste siano verificate sia per equazioni ellittiche che per altre più generali.

Dopo di ciò noi ci limiteremo principalmente alle equazioni astratte (1.1) con pochi ulteriori riferimenti alle equazioni differenziali alle derivate parziali.

Noi considereremo i seguenti problemi:

(i) Sviluppi asintotici per grandi valori di t delle soluzioni di

$$Lu = 0 \qquad t > 0 \tag{1.2}$$

come somma delle "soluzioni esponenziali". Completezza di queste soluzioni esponenziali.

(ii) Regolarità delle soluzioni di $Lu = f$.

(iii) Unicità del problema di Cauchy includendo il caso del problema di Cauchy all'infinito cioè : la soluzione nulla è la sola soluzione che tende a zero rapidamente all'infinito. Considereremo qui più generalmente funzioni u che soddisfano disuguaglianze del tipo

$$|Lu| \leq \varphi(t)\, |u| \tag{1.3}$$

dove $|\ \ |$ rappresentano la norma nello spazio di Banach. Noi otterremo anche limitazioni inferiori per le soluzioni in vari casi mostrando che le soluzioni non possono tendere a zero troppo rapidamente; queste limitazioni sono ottenute con considerazioni di convessità .

(iv) stabilità all'infinito; cioè le soluzioni di quadrato integrabile della (1.3) tendono a zero in modo esponenziale.

1.2.- Per vedere quali condizioni sull'operatore A è opportuno imporre, cominciamo col richiamare quale è la situazione in uno spazio di dimensioni finite; cioè per un sistema di un numero finito di equazioni differenziali ordinarie. Ogni operatore iA è allora limitato ed è il generatore di un semigruppo (nel caso attuale di un gruppo) limitato di operatori.

$$T(t) = e^{itA} \tag{1.4}$$

essendo $T(s+t) = T(s)T(t)$, $T(o) = I$

$$\frac{1}{i}\frac{dT}{dt} = A\,T(t) \tag{1.5}$$

e la soluzione di (1.1) con u assegnata, per fissare le idee, per $t = s$, è

$$u(t) = T(t-s)u(s) + i\int_s^t T(t-\tau)f(\tau)d\tau\,. \tag{1.6}$$

Le soluzioni dell'equazione omogenea $Lu = 0$ sono combinazioni lineari di polinomi esponenziali che chiameremo semplicemente soluzioni esponenziali $\chi(t) = e^{i\lambda_0 t}p(t)$, dove λ_0 è un autovalore di A e p è un polinomio in t i cui coefficienti sono vettori dello spazio;

$$p(t) = \varphi_m + it\,\varphi_{m-1} + \ldots\ldots + \frac{(it)^{m-1}}{(m-1)!}\varphi_1\,, \qquad \varphi_1 \neq 0\,. \tag{1.7}$$

Qui

$$(A-\lambda_0)\varphi_1 = 0,\ (A-\lambda_0)\varphi_j = \varphi_{j-1}, \qquad j = 2,\ldots\ldots,m; \tag{1.8}$$

m è l'indice della soluzione esponenziale. Ogni autovalore λ_0 è un polo della risolvente $R(\lambda) = R(\lambda, A) = (\lambda I - A)^{-1}$ che è una funzione meromorfa. Il massimo indice che una soluzione esponenziale $e^{i\lambda_0 t}p(t)$ può avere è semplicemente l'ordine del polo λ_0.

(Anche nel caso generale noi consideriamo soluzioni esponenziali, definite nello stesso modo, allora la (1.8) è automaticamente verificata, ed esse provengono dai poli dell'operatore risolvente $R(\lambda, A) = (\lambda I - A)^{-1})$.

Osserviamo che la norma dell'operatore $| T(t) |$ per valori positivi di t è limitata da una costante moltiplicata per $e^{\omega t}$ dove $-\omega$ è il minimo delle parti immaginarie degli autovalori λ_0.

E' chiaro che le soluzioni di $Lu = 0$ che sono di quadrato integrabile sull'asse positivo tendono a zero esponenzialmente per $t \to \infty$, infatti queste sono ovviamente combinazioni lineari di soluzioni esponenziali con $\operatorname{Im} \lambda_0 > 0$.

Una tale soluzione è $0(e^{-at})$ dove $a = \min \operatorname{Im} \lambda_0$ fra gli autovalori λ_0 tali che $\operatorname{Im} \lambda_0 > 0$. Si può anche dimostrare la tendenza a zero in modo esponenziale delle soluzioni di quadrato integrabile di alcuni sistemi con coefficienti variabili:

$$\frac{1}{i} \frac{du}{dt} - A(t)u = 0,$$

supposto che la matrice dei coefficienti $A(t)$ tenda con sufficiente rapidità a una matrice limite A quando $t \to \infty$. Allora noi scriveremo l'equazione sotto la forma

$$\frac{1}{i} \frac{du}{dt} - Au = (A(t) - A)u$$

o, più generalmente, sotto la forma(1.3), dove $\varphi(t)$ tende a zero opportunamente per $t \to \infty$. Senza alcuna condizione sul modo di tendere a zero di $A(t) - A$, o di $\varphi(t)$, non è vero in generale che soluzioni di quadrato integrabile tendono a zero esponenzialmente. Consideriamo il seguente semplice esempio: un'equazione in una sola variabile scalare

$$(1.9) \qquad \frac{dv}{dt} = \frac{-c}{1+t} v, \qquad t > 0$$

dove ovviamente, $A = 0$. Una soluzione $v = (1+t)^{-c}$ non tende a zero esponenzialmente ma è di quadrato integrabile se $c > \frac{1}{2}$.

La stessa funzione $v = (1+t)^{-c}$ è anche soluzione dell'equazione di ordine superiore.

$$(1.9') \qquad \left(\frac{d}{dt}\right)^k v = \frac{C}{(1+t)^k} v , \quad C = (-1)^k c(c+1) \ldots\ldots (c+k-1).$$

Questa equazione può essere scritta come un sistema del primo ordine per $u = (u_o, \ldots\ldots, u_{k-1})$ con $u_o = v$

$$(1.9'') \qquad \begin{cases} \dfrac{du_j}{dt} - u_{j+1} = 0 \\ \\ \dfrac{du_{k-1}}{dt} = \dfrac{C}{(1+t)^k} u_o \end{cases} \qquad j = 0, 1, \ldots, k-2.$$

Allora noi vediamo che le soluzioni di quadrato sommabile non tendono a zero esponenzialmente se $c > \frac{1}{2}$. Infatti in questo caso la corrispondente risolvente ha un polo di ordine k nell'origine; per poter concludere la tendenza a zero in modo esponenziale, occorre che $\varphi(t)$ (in questo caso $C(1+t)^{-k}$) sia limitata da cost. $(1+t)^{-k}$ e la costante sia sufficientemente piccola. Noi vedremo (capitolo 4) che la situazione è analoga per un operatore generale in uno spazio di Banach.

1.3. - Per trattare i casi di dimensione infinita noi assumeremo generalmente che A sia un operatore chiuso e che la sua risolvente $R(\lambda) = R(\lambda, A) = (\lambda I - A)^{-1}$ sia una funzione meromorfa in una regione del piano complesso λ. In molti problemi collegati con equazioni ellittiche la seguente proprietà è soddisfatta: $R(\lambda)$ è regolare in un "doppio settore"

(1.10) $$|\arg \pm \lambda| \leq \delta, \qquad |\lambda| \geq N$$

e soddisfa la condizione

(1.11) $$|\lambda R(\lambda)| \leq c;$$

inoltre sull'asse reale, per $|\lambda| \to \infty$, si ha:

(1.12) $$|\lambda \frac{d}{d\lambda} R(\lambda)| = 0 \left(\frac{1}{|\lambda|}\right)$$

Consideriamo alcuni esempi di equazioni differenziali alle derivate parziali in un cilindro e mostriamo come si possano dedurre le proprietà richieste per la risolvente. Osserviamo dapprima che è spesso molto utile considerare $u(t)$ appartenente ad uno spazio di Banach X mentre $\frac{du}{dt}$ e Au appartengono ad un altro spazio di Banach Y; comunque, per semplicità, noi lavoreremo con un unico spazio di Banach Y la cui norma indicheremo con $|\ |$.

1.4.-Nel primo esempio che considereremo non ci sono soluzioni esponenziali diverse da zero.

Esempio 1. - Lo spazio di Banach è quello delle funzioni continue nell'intervallo $0 \leq x \leq 1$; A è l'operatore $\frac{i}{\alpha} \frac{\partial}{\partial x}$, dove α è una costante; A agisce sulle funzioni di C^1, che si annullano per $x = 0$. Allora $R(\lambda)$ è uguale a $(\lambda - \frac{i}{\alpha} \frac{\partial}{\partial x})^{-1}$ ed esiste per tutti i valori complessi di λ.

$$R(\lambda) f = i \alpha \int_0^x e^{-i\lambda\alpha(x-y)} f(y) dy .$$

Si verifica facilmente che, per una opportuna costante C , si ha

$$|R(\lambda)| \leq c \frac{e^{\operatorname{Im}\lambda\alpha}}{\operatorname{Im}\lambda\alpha} \quad \text{se} \quad \operatorname{Im}\lambda\alpha > 0$$

$$|R(\lambda)| \leq c\,(1 - \operatorname{Im}\lambda\alpha)^{-1} \quad \text{se} \quad \operatorname{Im}\lambda\alpha \leq 0$$

Consideriamo ora l'equazione per $u(x,t)$

$$Lu = \frac{1}{i}\frac{du}{dt} - \frac{i}{\alpha}\frac{\partial u}{\partial x} = 0 \tag{1.13}$$

in $0 \leq t \leq T$, $0 \leq x \leq 1$. Se $\operatorname{Im}\alpha \neq 0$ l'equazione (1.13) è ellittica e ha la sola soluzione $u = 0$. Se $\operatorname{Im}\alpha = 0$ supponiamo $\alpha > 0$;in modo che iL risulta essere una derivata direzionale nel piano (x,t); allora la soluzione si annulla per $t > x\alpha$ ma non nécessariamente per $t < x\alpha$. D'altra parte $u = 0$ è la sola soluzione che si annulla per $t = 0$.

1.5. Esempio 2. - Consideriamo un'equazione ellittica di ordine $2m$ in un cilindro avente l'asse t come generatrice e un dominio limitato $\mathcal{D}$ nello spazio Euclideo n-dimensionale come base:

(poniamo $D_x = (\frac{1}{i}\frac{\partial}{\partial x_1}, \ldots\ldots, \frac{1}{i}\frac{\partial}{\partial x_n})$, $D_t = \frac{1}{i}\frac{\partial}{\partial t}$):

$$\mathcal{A}(x; D_x, D_t)\, u = D_t^{2m} u + \sum_{0}^{2n-1} A_{2m-j} D_t^j u = f \tag{1.14}$$

dove gli A_{2m-j} sono operatori differenziali nelle variabili x di ordine $\leq 2m-j$ con coefficienti indipendenti da t . Supponiamo che u soddisfi condizioni al contorno di ordine minore di $2m$ "coercitive" sulla superficie laterale del cilindro con coefficienti indipendenti da t della forma

(1.15) $$B_j(x, D_x)u = 0 \qquad j=1, \ldots, m.$$

Introducendo come nuove incognite le derivate rispetto a t

$$u_j = D_t^j u, \qquad j=0, \ldots, 2m-1,$$

possiamo scrivere la (1.14) come un sistema del primo ordine in t

(1.16) $$\begin{array}{ll} D_t u_j - u_{j+1} = 0 & j=0, \ldots, 2m-2 \\ \cdots\cdots & \\ D_t u_{2m-1} + \sum_{0}^{2m-1} A_{2m-j} u_j = f & \end{array}$$

Le proprietà (1.11), (1.12) per la risolvente dell' operatore che ne risulta si ottengono dalla seguente nota proprietà per le soluzioni del problema (1.14), (1.15) :

(A) L'integrale della somma dei quadrati di tutte le derivate di u fino all'ordine $2m$ su una sezione $t_1 < t < t_2$ è maggiorato da una costante per l'integrale di $(|f|^2 + |u|^2)$ in una sezione più larga $t_1' < t < t_2'$, $t_1' < t_1$, $t_2' > t_2$. Da ciò segue :

Teorema 1.1. - Sia v una funzione definita in $\mathfrak{D}$ che soddisfi le condizioni al contorno (1.15), esistono allora tre costanti positive δ, C, N che dipendono soltanto dal problema di valori al contorno ellittico, tali che per λ nel "doppio settore" (1.10) : $|\arg \pm \lambda| \leq \delta$, $|\lambda| \geq N$, si ha

(1.17) $$\sum_{j=0}^{2m} |\lambda|^{2m-j} \|v\|_j < C \|\mathcal{A}(x; D_x, \lambda)v\|$$

Qui $\| \;\|$ rappresenta la norma di L_2 in $\mathfrak{D}$ e $\|v\|_j$ rappresenta la somma delle norme in L_2 di v e di tutte le sue derivate fino all'ordine j.

Per valori di λ reali nel "doppio settore" la disuguaglianza (1.17) si ottiene applicandola (A) alla funzione $u(x,t) = v(x)e^{i\lambda t}$ con $t'_1 = 0$, $t_1 = 1$, $t_2 = 2$, $t'_2 = 3$.

Consideriamo ora lo spazio di Banach Y dei vettori $U = (u_o, \ldots\ldots, u_{2m-1})$ completato rispetto alla norma

$$|U| = \sum_{j=o}^{2m-1} \| u_j \|_{2m-j-1} \tag{1.18}$$

e prendiamo come dominio dell'operatore A i vettori in C^∞, U dove u_o soddisfa le condizioni al contorno (1.15). Per ottenere informazioni sulla risolvente occorre invertire il sistema, dove $F = (f_o, \ldots, f_{2m-1})$

$$\begin{array}{ll} \lambda u_j - u_{j+1} = f_j & \\ \cdots\cdots\cdots & j=0,\ldots,2m-2. \\ \lambda u_{2m-1} + \Sigma A_{2m-j} u_j = f_{2m-1} & \end{array} \tag{1.19}$$

Supponiamo che sia $f_j = 0$ per $j < 2m-1$. Allora eliminando le u_j per $j > 0$ con l'osservazione che

$$u_j = \lambda^j u_o \tag{1.20}$$

troviamo

$$\mathcal{A}(x; D_x, \lambda) u_o = f_{2m-1} \tag{1.21}$$

Dalla (1.18) e (1.20) segue che per valori reali di λ, $|\lambda| \geq N$,

$$|\lambda U| \leq C |F|$$

Allora se noi indichiamo con S l'insieme dei vettori F con

$f_j = 0$, $j < 2m-1$, e con $R_S(\lambda)$ la restrizione di $R(\lambda)$ a S abbiamo per λ reale, $|\lambda| > N$,

(1.22) $$|\lambda R_S(\lambda)| \leq C$$

Poichè sostituendo D_t nell'operatore con $e^{i\theta} D_t$, dove θ è un numero piccolo, l'operatore rimane ellittico, otteniamo la maggiorazione (1.22) per $\arg \lambda = \pm \theta$. Allora segue la maggiorazione desiderata (1.11) per $R_S(\lambda)$ nel settore (1.10). Siccome $R_S(\lambda)$ è analitica in λ nel settore (1.10), segue facilmente la (1.12).

Inoltre segue dalla teoria delle equazioni ellittiche che $R(\lambda)$ è meromorfa nell'intero piano.

Consideriamo altri due semplici esempi connessi, i quali sebbene riguardano problemi non ellittici possono essere trattati per mezzo del teorema 1.1 applicato ad un operatore opportunamente modificato.

Esempio 3. - Di nuovo nel cilindro, consideriamo l'operatore

$$D_t u - A(x, D_x) u$$

dove A è un operatore di ordine $2m$, insieme con, per semplicità, le condizioni al contorno di Dirichlet (1.15).

Assumiamo inoltre che, se $A'(x, D_x)$ è la parte principale di A, in ogni punto x, si abbia

(1.24) $$\tau - A'(x, \xi) = 0, \qquad \text{per } \tau, \xi \text{ reali solo se } \tau = 0, \ \xi = 0.$$

Allora in particolare $D_t^{2m} \pm A(x, D_x)$ sono operatori ellittici. In conseguenza, se lo spazio di Banach è L_2, troviamo di nuovo che la risolvente $R(\lambda)$ soddisfa la (1.11) e la (1.12) nel doppio settore (1.10).

Questo si dimostra come prima applicando il teorema 1.1 agli operatori $D_t^{2m} \pm A(x, D_x)$.

Esempio 4. - Nel solito cilindro consideriamo l'operatore

$$D_t^2 u - \Delta_x^2 u$$

il quale agisca sulle funzioni che si annullano con le derivate prime sulla superficie laterale del cilindro. Scrivendo l'operatore come un sistema del primo ordine per due funzioni e usando il fatto che $D_t^4 + \Delta_x^2$ è ellittico, così che il Teorema 1.1 può essere applicato, noi troviamo di nuovo che la (1.11) e la (1.12) sono soddisfatte per $R_S(\lambda)$, cioè per $R(\lambda)$ ristretto ai vettori (0, f).

1.6. Noi faremo spesso uso di condizioni come (1.11), (1.12); come abbiamo visto queste possono non valere in pratica per $R(\lambda)$ ma solo per $R_S(\lambda)$, cioè per $R(\lambda)$ ristretta ad un opportuno sottospazio di Y. Molti dei risultati possono essere estesi ai casi nei quali le (1.11), (1.12) valgono soltanto per $R_S(\lambda)$, ma per semplicità noi non prenderemo in considerazione questo raffinamento (cf. [2]).

Osserviamo che il problema (iv) sulla tendenza a zero di tipo esponenziale può essere usato per dimostrare che lo spazio delle soluzioni di quadrato sommabile di alcuni problemi al contorno ellittici omogenei, i cui coefficienti tendono rapidamente a valori limiti quando $t \to +\infty$, ha dimensione finita. Si potrebbe sperare che questo sia il caso generale per un problema ellittico uniforme ma è possibile dare un semplice contro esempio per mezzo di un esempio di Plis [12]. Plis ha costruito un operatore ellittico lineare T con coefficienti principali reali per il quale vi è una soluzione non banale v con supporto nella sfera unitaria. Sia ora Γ il cilindro sopra la sfera e sia L un operatore ellittico in Γ con coefficienti periodici (con periodo 2π) nella direzione x_{n+1} della generatrice in modo tale che $L = T$ nella sfera unitaria col centro nell'origine. Allora per ogni intero j poniamo $v_j = v(x_1, \dots, x_n, x_{n+1} - 2\pi j)$ dove v è la soluzione costruita da Plis. Le soluzioni v_j hanno dati di Cauchy nulli e sono linearmente indipendenti.

Da ora in poi noi ci limiteremo principalmente alla teoria astratta ma converrà tener presente questi esempi. Noi daremo dimostrazioni quasi complete perchè le tecniche usate, che forse hanno interesse maggiore che i risultati stessi, non sono molto complicate - sebbene qualche volta un poco artificiose. Noi faremo costante uso della trasformata di Fourier, della teoria elementare delle funzioni di variabile complessa, in particolare del teorema di Phragmén-Lindelöf e del teorema di Paley-Wiener.

Capitolo 2

Sviluppi in serie di soluzioni esponenziali.

2.1. In questo capitolo noi considereremo alcuni risultati abbastanza semplici per l'equazione

$$(2.1) \qquad Lu = \left(\frac{1}{i}\frac{\partial}{\partial t} - A\right)u = 0 \qquad t > 0$$

i quali illustreranno l'uso della trasformata di Fourier e della teoria delle funzioni di variabile complessa. Noi supporremo che $|u(t)|$ sia integrabile nell'intervallo $0 < t < \infty$. Se noi poniamo $u(t) = 0$ per $t < 0$ e consideriamo la trasformata di Fourier di u

$$\hat{u}(\lambda) = \frac{1}{\sqrt{2\pi}} \int_0^{\infty} e^{-i\lambda t} u(t)dt$$

allora troveremo, prendendo la trasformata di Fourier in (2.1)

$$(2.2) \qquad (\lambda - A)\hat{u}(\lambda) = \frac{1}{i\sqrt{2\pi}} u(0)$$

o

$$(2.3) \qquad \hat{u}(\lambda) = \frac{1}{i\sqrt{2\pi}} R(\lambda)u(0)$$

per quei valori reali di λ per i quali $R(\lambda)$ esiste. Poichè $u(t)$ si annulla per $t < 0$ la funzione $\hat{u}(\lambda)$ può essere estesa in una funzione analitica nel semipiano $\operatorname{Im}\lambda < 0$ la quale è continua in $\operatorname{Im}\lambda < 0$. Se $R(\lambda)$ è regolare in qualche regione del semipiano superiore la quale tocchi l'asse reale λ (come (1.10)) allora la (2.3) può essere usata per estendere analiticamente $\hat{u}(\lambda)$ in questa regione.

Noi cominciamo con un semplice risultato per il problema di Cauchy finito, essenzialmente dovuto a Lyubič [11]

. Per uniformità noi formuliamo questo risultato assumendo $u(T) = 0$ e proviamo che $u(t) = 0$ per $t < T$.

Teorema 2.1 : Sia $u(t)$ una soluzione di (2.1) per $0 \leq t \leq T$ con $u(T) = 0$. Supponiamo che ci sia un arco di Jordan semplice γ il quale vada all'infinito e rimanga in un angolo chiuso nel semipiano aperto $\operatorname{Im}\lambda > 0$, nel quale $R(\lambda)$ è definito e soddisfa $R(\lambda) = 0(e^{\alpha \operatorname{Im}\lambda})$ per qualche costante $\alpha \geq 0$. Se $T > \alpha$ allora $u(t) = 0$ per $t > \alpha$.

Il risultato è molto preciso in quanto $u(t)$ non si annulla necessariamente per $t < \alpha$ come possiamo vedere nell'esempio 1 della sezione 1 quando α è reale e positivo. Dim. : Estendendo $u(t)$ col valore zero per t fuori dall'intervallo $0 \leq t \leq T$ e prendendo la trasformata di Fourier, come prima, troviamo che $\hat{u}(\lambda)$ è una funzione vettoriale intera che soddisfa la (2.2). Poichè sulla curva γ la (2.3) è verificata, si ha

$$|\hat{u}(\lambda)| = 0(e^{\alpha \operatorname{Im}\lambda})$$

su γ. D'altra parte sull'asse reale $|\hat{u}(\lambda)|$ è limitata.

Per il teorema di Phragmén-Lindelöf concludiamo che

$$|\hat{u}(\lambda)| = 0(e^{\alpha \operatorname{Im}\lambda})$$

nell'intero semipiano superiore. Segue dal teorema di Paley-Wiener che $u(t) = 0$ per $t > \alpha$.

2.2. Consideriamo ora alcuni risultati connessi con i problemi (i), (ii), (iii). Formuliamo le seguenti ipotesi

(a) $R(\lambda)$ è regolare nelle due regioni angolari

(2.4) $$0 \leq \arg(\lambda - N) \leq \theta, \qquad 0 \leq \pi - \arg(\lambda + N) \leq \theta$$

con qualche costante positiva N per $\theta < \frac{\pi}{2}$, e soddisfa la condizione

(2.4') $$|R(\lambda)| = 0\,(e^{\alpha \operatorname{sen}\theta\, |\lambda|}) \qquad \text{per } |\lambda| \to \infty$$

(b) $R(\lambda)$ è una funzione meromorfa nel semipiano chiuso $\operatorname{Im}\lambda \geq 0$. Indichiamo con $\lambda_1, \lambda_2, \ldots\ldots\ldots$ i poli nell'insieme $\operatorname{Im}\lambda > 0$, numerati in ordine crescente delle parti immaginarie.

Teorema 2.2 : Nella ipotesi (a) $u(t)$ può essere esteso in una funzione analitica di $t = \sigma + i\tau$ (con valori in Y) nella regione angolare

(2.5) $$|\arg(t - \alpha)| < \theta .$$

Assumendo che anche l'ipotesi (b) valga e indicando con $u_j = e^{i\lambda_j t}\, p_j(t)$ il residuo di $e^{i\lambda t} R(\lambda)\, u(0)$ nel polo λ_j, allora

$$\sum_1^{\infty} u_k(t)$$

rappresenta uno sviluppo asintotico per $u(t)$. Inoltre se $\lambda_1, \ldots\ldots, \lambda_m$, sono i poli contenuti nella striscia $0 < \operatorname{Im}\lambda < a$, in modo che $\operatorname{Im}\lambda_m < a - \varepsilon$ per qualche $\varepsilon > 0$, e se $\mu = \sigma - \alpha + \tau \operatorname{cotg}\theta$ allora

(2.6) $$\left| u(t) - \sum_1^m u_k(t) \right| \leq \text{costante}\ |u(0)|\ \frac{e^{N|\tau|}}{\mu}\ e^{-(a-\varepsilon)\mu}$$

(nel settore (2.5)).

Nell'esempio 1 della sezione 1 con $\alpha > 0$ tutte le condizioni del teorema sono soddisfatte; infatti non vi sono soluzioni esponenziali, e di più per $t < \alpha$ la soluzione non è necessariamente analitica e la (2.6) non vale.

La dimostrazione del teorema segue la via diretta. La relazione (2.3) serve ad estendere $\hat{u}(\lambda)$ in una funzione analitica nella regione angolare (2.4). Noi possiamo scrivere

$$u(t) = \frac{1}{2\pi i}\int_{-\infty}^{-c} e^{i\lambda t} R(\lambda)u(0)\,d\lambda + \frac{1}{2\pi i}\int_{c}^{\infty} e^{i\lambda t} R(\lambda)u(0)\,d\lambda$$

$$+ \frac{1}{\sqrt{2\pi}}\int_{-c}^{c} e^{i\lambda t}\,\hat{u}(\lambda)\,d\lambda$$

I primi due integrali possono essere calcolati lungo i lati obbliqui degli angoli definiti da (2.4) (questo può essere giustificato facilmente con l'uso di un argomento che fa uso del teorema di Phragmén-Lindelöf); questi integrali convengono assolutamente per t nella regione (2.5) (per esempio l'integrale $\int_{N}^{N+e^{i\theta x}} e^{i\lambda t} R(\lambda)\,u(0)\,d\lambda$ converge assolutamente per t nel semipiano $0 < \arg(t-\alpha) + \theta < \pi$) mentre l'ultimo è una funzione intera di t. Così il primo asserto del teorema è provato.

Nel caso che l'ipotesi (b) valga noi possiamo usare la (2.3) per estendere $\hat{u}(\lambda)$ in una funzione meromorfa nel semipiano $\operatorname{Im}\lambda \geq 0$ e siccome $|\hat{u}(\lambda)|$ è limitato sull'asse reale non vi sono ivi poli. Allora possiamo sostituire il terzo integrale con l'integrale di $\frac{1}{2\pi i} e^{i\lambda t} R(\lambda)u(0)$ esteso a una curva che congiunge $-N$ con N e spostarla di poco dall'asse reale in modo da evitare i poli reali di $R(\lambda)$. Il contorno che ne risulta, costituito dai lati degli angoli (2.4) e della curva che congiunge $-N$ con N può essere spostato verso l'alto in modo che l'integrale dia i residui $u_j(t)$ e la disuguaglianza (2.6) è facilmente dimostrata.

Un simile argomento, integrando lungo i lati obliqui relativi alla trasformata di Fourier di u, dimostra che se $R(\lambda)$ è regolare nella regione

$$|\arg \pm \lambda| \leq \theta \leq \frac{\pi}{2}, \qquad |\lambda| \geq N$$

e soddisfa la relazione

$$|R(\lambda)| = 0(e^{\alpha |\mathrm{Im}\lambda|}), \qquad \alpha \geq 0$$

allora ogni soluzione di $Lu = f$ appartenente alla classe C^1 con f analitica sull'intervallo $|t| \leq T$, è analitica nell'intervallo $|t| \leq T - \alpha$. Inversamente se le soluzioni di $Lu = f$ con f analitica sono analitiche allora si ottengono, dal teorema del grafico chiuso, condizioni necessarie che sono molto simili alle condizioni sufficienti.

2.3. E' interessante osservare che dalla (2.6) si possono dedurre sia limitazioni inferiori che limitazioni superiori per $|u(t)|$ usando un argomento già usato da Krein e Prozorovskaya [7]. Supponiamo che valga l'ipotesi (a) e che $R(\lambda)$ sia regolare sull'asse reale eccetto al più per un numero finito di poli. Sia φ un numero positivo $< \theta$. Allora se $t_0 > \alpha$ c'è un numero β, dipendente dalla soluzione e da t_0, tale che, per $t > t_0$,

$$(2.7) \qquad \begin{aligned} &|u(t)| \geq |u(o)|\, e^{-\beta t^{1/2\varphi}} && \text{se } \theta < \pi/2 \\ &|u(t)| \geq |u(o)|\, e^{-\beta t \log t} && \text{se } \theta = \pi/2 \end{aligned}$$

Noi daremo un'idea della dimostrazione nel caso $\theta = \frac{\pi}{2}$. Sia μ_0 un numero positivo $< t_0 - \alpha$ e poniamo $\delta = \frac{t_0 - \alpha - \mu_0}{T}$ per $t \geq t_0$. La disuguaglianza (2.1) segue dalla disuguaglianza di convessità:

$$(2.8)\quad |u(t_o)| \leq C_1 C_2^{t_o} |u(T)|^{\delta} |u(0)|^{1-\delta} \exp(C_3 t \log \frac{1}{\delta} \log \frac{1}{1-\delta})$$

fissando t_o e ponendo $T=t$. Per ipotesi la disuguaglianza (2.6) vale per $m = 0$ nella forma (qui $t = \sigma + i\tau$)

$$(2.6')\qquad |u(t)| \leq \text{costante } |u(0)| e^{N|\tau|}$$

se $\sigma - \alpha \geq \mu_o$.

Consideriamo la striscia $\mu_o + \alpha \leq \sigma \leq T + \alpha + \mu_o$ nella quale la (2.6') certamente vale. Possiamo anche asserire che sul lato destro della striscia

$$|u(t)| \leq \text{costante } |u(T)| e^{N|\tau|}$$

considerando T al posto dell'origine. Siamo allora in condizione di applicare il "teorema delle tre rette" (in una forma opportuna) alla funzione analitica $u(t)$ in una striscia e la disuguaglianza (2.8) ne risulta.

2.4. Prendiamo ora in considerazione la questione della completezza delle soluzioni esponenziali. Noi supporremo che le precedenti ipotesi (a), (b) valgano in modo che, per il teorema 2.2, ogni soluzione u di $Lu = 0$ sull'asse positivo possiede uno sviluppo asintotico come somma di soluzioni esponenziali $\sum u_j(t)$. Noi diremo che le soluzioni esponenziali sommabili sono complete fra le soluzioni integrabili per $t \geq a$ se si verifica quanto segue: sia u una soluzione dell'equazione $Lu = 0$ integrabile per $t > 0$ e $\sum u_j(t)$ sia il suo sviluppo asintotico; assegnate due costanti positive ε, C esiste una combinazione lineare finita $\psi(t)$ delle $u_j(t)$ e delle loro traslate $u_j(t + \text{costante})$ tale che

(2.9) $$|u(t) - \psi(t)| \leq \varepsilon e^{-Ct} \qquad \text{se} \quad t \geq a.$$

Per provare la completezza noi faremo uso della nozione di "Ordine": $R(\lambda)$ è di ordine finito $\omega \geq 0$ per $\mathrm{Im}\,\lambda \geq 0$ se per ogni $\varepsilon > 0$ esiste una successione di curve di Jordan differenziabili J_n, contenute nel semipiano $\mathrm{Im}\,\lambda > 0$ eccettuati i punti terminali che appartengono all'asse reale da parte opposta rispetto all'origine (con la distanza di J_n dall'origine tendente all'infinito), tali che i) $R(\lambda)$ esiste su J_n e soddisfa la disuguaglianza

$$|R(\lambda)| \leq e^{|\lambda|^{\omega+\varepsilon}}$$

ii) ω è il più piccolo numero non negativo soddisfacente questa proprietà.

Per un operatore ellittico in un cilindro come nell'esempio 2 Agmon [1] ha dimostrato che la risolvente corrispondente a tale operatore A sotto forma di sistema del primo ordine ha "ordine" $\leq n$; n essendo la dimensione della base del cilindro. Anche per l'operatore dell'esempio 3 egli ha dimostrato che $\omega \leq \frac{n}{2m}$

Noi consideriamo operatori che soddisfano la seguente condizione:

(c) Assumiamo che $R(\lambda)$ abbia "ordine" ω in $\mathrm{Im}\,\lambda \geq 0$ e che esistano archi semplici differenziabili non sovrapponentisi $\gamma_1, \ldots\ldots, \gamma_k$ uscenti da un punto dell'asse reale e d'altra parte appartenenti al semipiano $\mathrm{Im}\,\lambda > 0$, tali che ognuna delle $k + 1$ regioni nelle quali il semipiano è diviso da questi archi sia contenuta in un angolo con apertura $< \frac{\pi}{\omega}$.

Di più $R(\lambda)$ esista su ciascun γ_j per valori di $|\lambda|$ grandi e inoltre per qualche costante $\beta > 0$, si abbia

$$|R(\lambda)| = O(e^{\beta|\lambda|}) \qquad \text{quando} \quad |\lambda| \to \infty,\ \lambda \in \gamma_j.$$

<u>Teorema 2.3. : Se</u> $R(\lambda)$ <u>soddisfa le condizioni</u> (a), (b), (c) <u>allora le soluzioni esponenziali integrabili sono complete nel senso precisato sopra, sull'intervallo</u> $t \geq \alpha + \beta + \delta$ <u>per ogni</u> $\delta > 0$.

Come dimostra l'esempio 1 le soluzioni esponenziali possono non essere complete sull'intero intervallo $t > 0$. <u>Dim.</u> : (1) Sia u una soluzione di $Lu = 0$ sommabile sull'asse positivo e $\sum u_j$ ne sia il suo sviluppo asintotico. Dapprima dimostriamo che dato $\varepsilon > 0$ esiste una combinazione lineare finita $\varphi(t)$ delle u_j e delle loro traslate tale che

$$|u(\beta) - \varphi(\beta)| \leq \varepsilon ,$$

cioè $u(\beta)$ appartiene alla chiusura della varietà generata dalle u_j e dalle loro traslate. Per provare ciò è sufficiente mostrare che se h^* è un funzionale lineare continuo definito su Y il quale sia zero su tutte le $u_j(t)$ per tutti i valori di t, allora $h^*(u(\beta)) = 0$. Questo si dimostra di nuovo con l'ausilio della trasformata di Fourier $\hat{u}$; $\hat{u}(\lambda)$, data dalla (2.3) nel semipiano $\operatorname{Im}\lambda \geq 0$, è meromorfa nell'intero piano con poli in λ_j, gli stessi poli di $R(\lambda)$. La funzione $h^*(\hat{u}(\lambda))$ è allora una funzione scalare analitica nell'intero piano eccettuato al più nei poli λ_j. Comunque i coefficienti delle potenze negative di $(\lambda - \lambda_j)$ nello sviluppo di Laurent di $\hat{u}(\lambda)$ relativo al punto λ_j sono vettori che appartengono alla varietà generata da u_j e dalle sue traslate valutate nel punto $t = 0$ (cfr. [4] capitolo VII).

Di conseguenza h^* si annulla su questi coefficienti e $h^*(u(\lambda))$ è una funzione intera.

La funzione $h^*(\hat{u}(\lambda))$ è limitata nel semipiano $\operatorname{Im}\lambda \leq 0$ perchè tale è $|\hat{u}(\lambda)|$. Di più noi possiamo applicare il teorema di Phragmén-Lindelöf in ciascuno degli angoli nei quali le curve γ_j dividono il semipiano superiore, e possiamo concludere che $h^*(\hat{u}(\lambda)) = O(e^{\beta \operatorname{Im}\lambda})$ per $\operatorname{Im}\lambda \geq 0$.

Poichè $h^*(\hat{u}(\lambda))$ è la trasformata di Fourier di $u(t)$ segue usando il teorema di Paley-Wiener che $h^*(u(t)) = 0$ per $t \geq \beta$.

(2) Per completare la dimostrazione del teorema supponiamo $\operatorname{Im}\lambda_j > C$ per $j > m$ e consideriamo la funzione

$$v(t) = u(t) - \sum_1^m u_j(t).$$

Ovviamente lo sviluppo asintotico di $v(t)$ è $\sum_{j>m} u_j(t)$. Applicando il risultato ottenuto in (1) alla funzione $v(t)$ troviamo che, dato un $\varepsilon' > 0$, esiste una combinazione lineare finita $\varphi(t)$ delle u_j e delle loro traslate, $j > m$, tale che $|v(\beta) - \varphi(\beta)| \leq \varepsilon'$. Se noi ora applichiamo la (2.6) alla funzione $v(t) - \varphi(t)$ troviamo;

$$|v(t) - \varphi(t)| < \frac{\text{costante}}{\delta} |v(\beta) - \varphi(\beta)| \, e^{-Ct} \qquad \text{per } t \geq \beta + \alpha + \delta.$$

Combinando insieme questa disuguaglianza con la precedente otteniamo il risultato desiderato. Q.E.D.

Come illustrazione dell'uso del teorema 2.3 consideriamo l'operatore differenziale parabolico

$$L = \frac{1}{i} \frac{\partial}{\partial t} - \frac{1}{i} \Delta_x ,$$

nel solito cilindro con base in uno spazio n-dimensionale, applicato alle funzioni che si annullano sulla superficie laterale. Per il risultato di Agmon precedentemente citato la risolvente corrispondente all'operatore ha "ordine" $\leq n/2$. D'altra parte per ogni numero complesso σ di modulo uno che non sia puramente immaginario l'operatore

$$D_t - \frac{\sigma}{i} \Delta_x$$

soddisfa la condizione dell'esempio 3 e ne segue che su ogni raggio $\arg \lambda = \theta$, $\theta \neq \frac{\pi}{2}, \frac{3\pi}{2}$ la risolvente $R(\lambda)$ esiste per $|\lambda|$ sufficientemente grande e soddisfa le condizioni di (c) con $\beta = 0$.

Quindi le soluzioni esponenziali di $Lu = 0$, sommabili sono dense in ogni intervallo $t \geq \delta > 0$ nell'insieme di tutte la soluzioni sommabili sul semiasse $t > 0$. (Infatti si può dimostrare che tali soluzioni sono complete sull'asse $t \geq 0$, anche se Δ_x è sostituito da un qualunque operatore fortemente ellittico il quale agisca su funzioni che hanno dati di Dirichlet nulli sulla superficie laterale del cilindro).

2.5. Terminiamo questo capitolo con un risultato relativo alle soluzioni di $Lu = 0$ definite per tutti i valori di t, il quale risultato noi chiameremo un principio astratto di Weinstein. Per semplicità noi non lo presenteremo nella sua forma più generale. Quando applicato a certe equazioni differenziali a derivate parziali in un cilindro completo $-\infty < t < \infty$, esso asserisce che lo spazio delle soluzioni crescenti al più esponenzialmente all'infinito ha dimensione finita.

Teorema 2.4. Sia $u(t)$ una soluzione di $Lu = 0$ sull'intero asse t tale che, per qualche $a > 0$, $e^{-a|t|}|u(t)|$ sia sommabile. Supponiamo che $R(\lambda)$ sia una funzione meromorfa nella striscia $\Sigma : |\mathrm{Im}\,\lambda| \leq a$ dove essa soddisfi la disuguaglianza $|R(\lambda)| = 0(e^{k|\lambda|})$ per valori grandi di $|\lambda|$ e per qualche $k > 0$. Allora $u(t)$ è la somma di un numero finito di soluzioni esponenziali corrispondenti ai poli di $R(\lambda)$ nella striscia aperta $|\mathrm{Im}\,\lambda| < a$. Se ciascun polo nella striscia aperta ha molteplicità finita allora l'insieme delle soluzioni u per cui $e^{-a|t|}|u(t)|$ è sommabile ha dimensione finita.

Ricordiamo (cfr. [4], § VII. 3) che se λ_0 è un polo di $R(\lambda)$ di ordine τ allora la dimensione dello spazio degli zeri di $(\lambda I - A)^{\tau}$, se finita, è chiamata la molteplicità di λ_0.

Dim.: Poniamo

$$\hat{u}_{\pm}(\lambda) = \frac{1}{\sqrt{2\pi}} \int_0^{\pm\infty} e^{-i\lambda t} u(t)dt$$

Per l'ipotesi formulata su u, $\hat{u}_+(\lambda)$ è analitica nel semipiano $\mathrm{Im}\,\lambda < a$ ed è limitata e continua nella chiusura del detto semipiano, mentre $\hat{u}_-(\lambda)$ è analitica nel semipiano $\mathrm{Im}\,\lambda > -a$ ed è limitata e continua nella sua chiusura. Di più nel semipiano corrispondente abbiamo

$$(\lambda I - A)\hat{u}_{\pm}(\lambda) = \frac{1}{i\sqrt{2\pi}} u(o) ,$$

o

$$\hat{u}_{\pm}(\lambda) = \frac{1}{i\sqrt{2\pi}} R(\lambda) u(o)$$

dovunque $R(\lambda)$ è definito. Siccome $R(\lambda)$ è meromorfa nella striscia $|\mathrm{Im}\,\lambda| < a$ questa formula fornisce la comune estensione analitica di $\hat{u}_+(\lambda)$ e $\hat{u}_-(\lambda)$ in questa striscia come funzioni meromorfe così che $\hat{u}_+$ e $\hat{u}_-$ sono estensioni analitiche l'una dell'altra. Indichiamo con $w(\lambda)$ la funzione meromorfa nell'intero piano definita da queste. Gli unici poli di $w(\lambda)$ sono i poli $\lambda_1, \dots\dots, \lambda_m$ di $R(\lambda)$ nella striscia aperta $|\mathrm{Im}\,\lambda| < a$.

Sul bordo della striscia $w(\lambda)$ è limitata, infatti $w(\lambda) = o(1)$ per $|\lambda| \to \infty$ per il teorema di Riemann-Lebesgue. A causa delle ipotesi formulate su $R(\lambda)$ possiamo applicare un teorema di Phragmén-Lindelöf per dedurre che $|w(\lambda)| = o(1)$ per $|\lambda| \to \infty$ nella striscia. Se $R_j(\lambda)$

è la parte singolare dello sviluppo di Laurent di $R(\lambda)$ relativo al polo λ_j si vede che la funzione

$$w(\lambda) - \frac{1}{i\sqrt{2\pi}} \sum_j R_j(\lambda)\, u(0)$$

è una funzione intera limitata tendente a zero per $|\lambda| \to \infty$ sull'asse reale. Per il teorema di Liouville questa funzione è zero, per modo che

$$u_{-}(\lambda) = \frac{1}{i\sqrt{2\pi}} \sum R_j(\lambda)\, u(0) .$$

Prendendo la trasformata inversa di Fourier si trova formalmente che $u(t)$ è uguale alla somma dei residui di $e^{i\lambda t} R(\lambda)\, u(0)$ nei punti λ_j, cioè uguale a una somma di soluzioni esponenziali.

L. Nirenberg

Capitolo 3

Unicità per il problema di Cauchy e proprietà di convessità.

3.1. In questo capitolo noi studieremo il problema (iii) per le soluzioni delle disuguaglianze (1.3) - del resto parecchi risultati essendo già stati ottenuti nella precedente sezione. Il nostro scopo qui è quello d'illustrare l'uso della convessità per ottenere limiti inferiori per la soluzione quando $t \to +\infty$. Come semplice esempio abbiamo il seguente noto risultato dove lo spazio Y è uno spazio di Hilbert.

Teorema 3.1. Sia $u(t)$ una soluzione di

$$Lu = \frac{1}{i}\frac{du}{dt} - Au = 0 \tag{3.1}$$

dove $A = \gamma B$, γ essendo una costante e B un operatore simmetrico. Allora $\log |u(t)|$ è una funzione convessa di t.

Allora per $0 \le t_0 \le t$ troviamo

$$|u(t_0)| \le |u(o)|^{1-t_0/t} \, |u(t)|^{t_0/t}$$

o, un limite inferiore

$$|u(t)| > |u(o)| \left(\frac{|u(t_0)|}{|u(o)|}\right)^{t/t_0} \quad ;$$

quindi di qui la tendenza a zero per $|u(t)|$ è al più esponenziale.

Per dimostrare il teorema basta osservare, derivando, che la derivata seconda di $\log |u(t)|^2$ è non negativa.

E' utile nella pratica avere tali risultati non soltanto per le soluzioni dell'equazione $Lu = 0$ ma anche per le soluzioni della disequazione (1.3) :

$|Lu| \le \varphi(t) |u|$. A questo proposito in [8] Lax ha dato una dimostrazione del risultato noto che ogni soluzione dell equazione a derivate parziali $\Delta u + \sum a_i \frac{\partial u}{\partial x_i} + au = 0$ la quale si annulli più rapidamente che ogni potenza nell'origine è identicamente zero. Con un ingegnoso ragionamento egli riduce la dimostrazione al seguente teorema elementare (dove Y è uno spazio di Hilbert).

Teorema 3.2. Sia $u(t)$ una soluzione della disuguaglianza

$$|Lu| \leq \varphi(t)\,|u| \qquad t \geq 0 \tag{3.2}$$

dove $u(t)$ e $\frac{du}{dt}$ sono di quadrato integrabile. Assumiamo che vi sia una successione di linee nel piano complesso λ parallele all'asse reale; $\operatorname{Im}\lambda = a_n$ con $a_n \longrightarrow \infty$ sulle quali $|R(\lambda)|$ è uniformemente limitata da una costante M. Se $\varphi(t) \leq c < M^{-1}$ e se $|u(t)| = 0\,(e^{at})$ per ogni valore reale a, allora $u \equiv 0$.

Il teorema si dimostra facilmente con uso delle trasformate di Fourier e del teorema di Parseval. Sia $\zeta(t) \geq 0$ una funzione monotona di t di C^{∞} la quale si annulli per $t \leq 0$ e sia uguale ad uno per $t \geq \tau$. Poniamo $v(t) = e^{a_n t}\zeta(t)\,u(t)$, con $v(t) = 0$ per $t < 0$, e posto

$$(L + i a_n)v = e^{a_n t}(\zeta Lu + uD_t\zeta) = f,$$

prendendo le trasformate di Fourier, abbiamo, per λ reale,

$$(\lambda + i\,a_n - A)\hat{v} = \hat{f}. \tag{3.3}$$

Allora per l'ipotesi su $R(\lambda)$ abbiamo ancora per λ reale

$$|\hat{v}(\lambda)| \leq M\,|\hat{f}(\lambda)| \tag{3.4}$$

e dal teorema di Parseval segue che, per un'opportuna costante C indipendente da n :

$$\int_0^\infty |v(t)|^2 dt \leq M^2 \int_0^\infty |f(t)|^2 dt$$

$$\leq M^2 \int_0^\tau |f|^2 dt + M^2 \int_\tau^\infty |e^{a_n t} Lu|^2 dt$$

$$\leq M^2 C e^{2a_n \tau} + c^2 M^2 \int_\tau^\infty e^{2a_n t} |u|^2 dt$$

a causa dell'ipotesi fatte.

Limitando l'integrazione a primo membro all'intervallo: $t > \tau$ dove $v = e^{a_n t} u$, troviamo, poichè $cM < 1$,

$$(1 - c^2 M^2) \int_\tau^\infty e^{2a_n t} |u|^2 dt \leq M^2 C e^{2a_n \tau}.$$

Questa disuguaglianza vale per tutti gli n, e ne segue di conseguenza che $u(t) = 0$ per $t > \tau$, o $u \equiv 0$ poichè τ è arbitrario.

Lax [8] ha anche dato un esempio per mostrare che la condizione $cM < 1$ è essenziale; in questo esempio iA è autoaggiunto, $M = \frac{1}{2}$, e u è una soluzione non banale della disuguaglianza $|Lu| \leq (\frac{1}{2} + \varepsilon) |u|$ (per un dato ε) la quale tende a zero come e^{-at^2} per qualche costante a.

3.2. I teoremi precedenti sono molto particolari; per esempio, nel secondo, se l'operatore iA è autoaggiunto, allora la condizione del teorema richiede che il complemento dello spettro all'asse reale non sia limitato superiormente su questo. E' naturalmente interessante ottenere estensioni di questi risultati che possono essere utili in pratica, e i risultati successivi di questa sezione possono essere considerati come generalizzazioni di essi. In questa sezione noi descriveremo soltanto due semplici generalizzazioni.

Consideriamo dapprima il teorema 3.2, noi cercheremo di avere un risultato analogo che sia valido per ogni operatore A con iA auto-aggiunto. In questo caso ogni segmento $\mathrm{Im}\,\lambda$ = costante nel piano comples-so λ può contenere punti dello spettro di A che appartengono all'asse immaginario. Assumendo ancora che Y sia uno spazio di Hilbert noi otterremo un teorema di unicità per il problema di Cauchy finito sotto le seguenti ipotesi:

(H) Consideriamo una successione di linee $\mathrm{Im}\,\lambda = a_n$, $a_n \to \infty$. Su ciascuna retta della successione $R(\lambda)$ esista fuori di un segmento di lunghezza s e abbia norma limitata da M. Nell'ipotesi (H) la posizione del segmento di lunghezza s può variare per ogni retta della successione.

<u>Teorema</u> 3.3. - <u>Sia</u> u <u>una soluzione della disequazione</u>

$$|Lu| \leq \varphi(t)\,|u|$$

<u>sull'intervallo</u> $0 \leq t \leq T$ <u>con</u> $u(T) = 0$ <u>e supponiamo che valga la</u> (H). <u>Esiste una costante</u> c <u>tale che se</u> $\varphi(t) < c$ <u>allora</u> $u \equiv 0$.

La dimostrazione segue da vicino quella del teorema 3.2; con $u(t) = 0$ per $t > T$ consideriamo $v(t) = e^{a_n t}\,\zeta(t)\,u(t)$, $(L + i a_n)v = f(t)$ e otteniamo la (3.3) come prima. La disuguaglianza (3.4) allora vale per tutti i valori reali di λ nel complemento I di un intervallo di lunghezza s per modo che

$$\int_I |\hat{v}(\lambda)|^2 d\lambda \leq M^2 \int_{-\infty}^{\infty} |\hat{f}(\lambda)|^2 d\lambda .$$

Poichè $v(t)$ ha il supporto nell'intervallo $0 \leq t \leq T$ possiamo mostrare che $\hat{v}(\lambda)$ ha la proprietà che su ogni retta $\mathrm{Im}\,\lambda = a$ la norma in L_2 di $|\hat{v}(\lambda)|$ è limitata da $e^{|a|T}$ moltiplicato per la norma in L_2 calcolata sull'asse reale. Poichè $\hat{v}(\lambda)$ è una funzione intera non è difficile dimostrare (per esempio, per assurdo) che c'è una costante k tale che

$$\int_{-\infty}^{\infty} |\hat{v}(\lambda)|^2 d\lambda \leq k \int_I |\hat{v}(\lambda)|^2 d\lambda .$$

Conseguentemente abbiamo

$$\int_{-\infty}^{\infty} |\hat{v}(\lambda)|^2 d\lambda \leq k M^2 \int_{-\infty}^{\infty} |\hat{f}|^2 d\lambda ;$$

dopo di che si procede come per il teorema 3.2.

Nella stessa situazione del teorema 3.2 sarebbe desiderabile ottenere anche limitazioni inferiori per le soluzioni della (3.2). In condizioni più onerose possiamo ottenere limitazioni inferiori per

$$\int_t^{t+\rho} |u(\tau)|^2 d\tau$$

per ogni $\rho > 0$ invece che per $|u(t)|$.

Teorema 3.4. Assumiamo che su ogni retta $\mathrm{Im}\,\lambda$ = costante ≥ 0 ci sia un intervallo di lunghezza s fuori del quale $R(\lambda)$ esiste e sia limitata in norma da una costante M fissa. Esiste una costante c tale che se $\varphi(t) \leq c$ allora ogni soluzione della (3.2) soddisfa la disuguaglianza

$$\int_t^{t+\rho} |u(\tau)|^2 d\tau \geq K_o K_1^t e^{-\beta\mu^t}$$

dove K_o, K_1, μ sono costanti fisse e β è una costante dipendente dalla soluzione.

La dimostrazione è basata su un argomento di convessità ed è troppo complicata per essere data qui.

Osserviamo che il metodo della dimostrazione del teorema 3.1 può essere anche trasportato all'equazione con coefficienti variabili (ancora in uno spazio di Hilbert con prodotto scalare (,))

$$\frac{du}{dt} - B(t)u = 0 \,. \tag{3.5}$$

Ci limitiamo ad enunciare il risultato. Per ogni t, $B(t)$ sia un operatore chiuso definito in un insieme denso dello spazio ed $u(t)$ appartenga sia al dominio di $B^*(t)$ che a quello di B. Assumiamo inoltre che esso dipenda regolarmente da t e che sia quasi autoaggiunto. Noi esprimiamo queste condizioni sotto la forma: esistono due costanti k, c tali che per ogni soluzione $u(t)$ di classe C^2 valga la seguente relazione

$$\operatorname{Re} \frac{d}{dt} (B(t)u(t), u(t)) \geq \frac{1}{2} |(B+B^*)u|^2 + \\ + c \operatorname{Re} ((B-k)u, u). \tag{3.5'}$$

Sotto queste condizioni se u è una soluzione di (3.5) di classe C^2 allora $\log |e^{-kt} u(t)|$ è una funzione convessa della variabile $\tau = e^{ct}$. La dimostrazione è analoga a quella del teorema 3.1 in quanto si dimostra che la derivata seconda di $\log|e^{-kt}u(t)|$ rispetto a τ è non negativa.

3.3. Ci occupiamo ora di un teorema di convessità nello spazio di Banach Y per le soluzioni della (3.2) sotto opportune condizioni sull'operatore A. Il risultato che noi presentiamo è una generalizzazione di un risultato recente di Cohen e Lees [3]. Faremo uso del "teorema delle tre rette" di Hadamard per funzioni analitiche in una striscia. L'operatore A sia della forma

$$A = A_\alpha = e^{-i\alpha} A_0 \qquad 0 \leq \alpha < \pi \quad , \tag{3.6}$$

dove iA_0 è un operatore lineare chiuso ed è un generatore infinitesimale di un gruppo fortemente continuo $T(t)$ di operatori. Assumeremo che gli operatori $T(t)$ siano uniformemente limitati:

$$|T(t)| \leq K \tag{3.7}$$

sebbene analoghi risultati valgano anche nel caso che $|T(t)| \leq K e^{\omega|t|}$. (Nel caso che Y sia uno spazio di Hilbert, un operatore autoaggiunto A_0 soddisfa certamente la (3.7))

Dapprima noi daremo una semplice estensione del teorema 3.1. Questa è ottenuta estendendo analiticamente la soluzione nel campo complesso. Consideriamo questa estensione anche per l'equazione inomogenea

$$\frac{du}{dt} - i A_\alpha u = i f . \tag{3.8}$$

Nel caso $\alpha = 0$, iA_0 essendo generatore di $T(t)$, abbiamo la formula (1.6)

$$u(s+r) = T(r)u(s) + i \int_0^r T(\lambda) f(t-\lambda) d\lambda \quad .$$

Per $\alpha \neq 0$ c'è una formula di rappresentazione simile nel caso che f sia olomorfa per valori complessi di $t = \tau + i\sigma$ in una striscia $\Sigma = \Sigma(a, b; \alpha)$ che è data da tutti valori complessi di t della forma $t = s + re^{i\alpha}$, $a < s < b$, $-\infty < r < \infty$; noi penseremo (s, r) come coordinate oblique nella striscia. <u>La seguente formula fornisce una estensione olomorfa di</u> u <u>in</u> Σ:

$$(3.9) \qquad u(t) = u(s+re^{i\alpha}) = T(r)u(s) + ie^{i\alpha} \int_0^r T(\lambda) f(t-\lambda e^{i\alpha}) d\lambda \quad .$$

Questo si verifica con un facile calcolo.

Nel caso $f=0$ noi vediamo dalla (3.7) che $|u(t)| \leq K |u(a)|$, $|u(t)| \leq K |u(b)|$ valgono rispettivamente nel lato sinistro e destro della striscia. Applicando il "teorema delle tre rette" di Hadamard otteniamo il seguente

<u>Teorema</u> 3.1' : <u>Se vale la</u> (3.7), <u>ogni soluzione di</u> $Lu = 0$ <u>su</u> (a, b) <u>soddisfa la disuguaglianza di convessità</u>

$$|u(t)| \leq K |u(a)|^{\frac{b-t}{b-a}} \cdot |u(b)|^{\frac{t-a}{b-a}} \qquad a < t < b$$

Supponiamo ora che iA_0 invece di essere il generatore infinitesimale di un gruppo sia il generatore infinitesimale di un semigruppo $T(t)$ per $t \geq 0$ fortemente continuo. La formula (3.9) dà ancora un estensione analitica di una soluzione di $Lu = 0$ su (a, b),

$$(3.10) \qquad u(s+re^{i\alpha}) = T(r)\; u(s)$$

in una semistriscia $r > 0$. Allora vale il seguente teorema sulla continuazione unica all'infinito (assumendo $\alpha > 0$).

Teorema 3.5. *Se* iA_0 *è un generatore infinitesimale di un semigruppo* $T(t)$ *fortemente continuo e se una soluzione* u *di* $Lu = 0$ *per* $t > 0$ *soddisfa la disuguaglianza* $|u(t)| = 0(e^{-\varepsilon t^{\pi/\alpha}})$ *per qualche* $\varepsilon > 0$, *allora* $u \equiv 0$.

Dim. - Siccome la norma di $T(t)$ cresce al più esponenzialmente noi possiamo sostituire $T(t)$ con $T(t)e^{-ct}$, dove $c > 0$ e conseguentemente u con $u\,e^{-\gamma t}$ ($\gamma = ce^{-i\alpha}$).
Questa soluzione soddisfa ancora alle stesse ipotesi così che possiamo ridurci al caso che $|T(t)| \leq K$. La (3.10) fornisce allora un'estensione analitica di u, uniformemente limitata, nell'angolo $0 < \arg t < \alpha$, la quale è fortemente continua nella chiusura dell'angolo e soddisfa la condizione $|u(t)| = 0(e^{-\varepsilon t^{\pi/\alpha}})$ sull'asse reale positivo.

Il cambiamento di variabile $z = t^{\pi/\alpha}$ trasforma l'angolo nel semipiano superiore del piano z; ponendo $u(z^{\alpha/\pi}) = v(z)$ vediamo che la norma di $v(z)$ nel semipiano $\mathrm{Im} z \geq 0$ è limitata mentre sull'asse reale positivo si ha $|v(z)| = 0(e^{-\varepsilon z})$. Allora la funzione analitica $e^{\frac{\varepsilon}{2} z}\, v(z)$ soddisfa la condizione $|v(z)| = 0(e^{-\frac{\varepsilon}{2}|z|})$ sull'asse reale. Per un classico teorema di Carlson (cfr. [13] p. 185) concludiamo che $v = 0$ e quindi anche $u \equiv 0$.

3.4. Le limitazioni inferiori per le soluzioni della (3.2) sono basate su disuguaglianze per una soluzione dell'equazione non omogenea

$$(3.11) \qquad \frac{du}{dt} - i\,A_\alpha = if\,, \qquad a < t < b\,,$$

per mezzo dei suoi valori agli estremi dell'intervallo (f e u sono supposte fortemente continue nell'intervallo chiuso $[a, b]$). Noi continueremo a supporre valida la (3.7).

Teorema 3.6. Esiste una costante fissa tale che ogni soluzione della (3.11) soddisfa la disuguaglianza

$$(3.12) \qquad |u(t)| \leq C\,K \left\{ |u(a)| + |u(b)| + \max_{a\leq t\leq b} \int_a^b |f(x)|\,(i+\log\frac{b-a}{|x-t|})dx \right\}.$$

Prima di dimostrare il teorema osserviamo che se lo spazio considerato Y è uno spazio di Hilbert e se iA è autoaggiunto allora vale una disuguaglianza più forte (e può essere provata in modo più semplice): per $a \leq t \leq b$

$$|u(t)|^2 \leq 2\,|u(a)|^2 + 2\,|u(b)|^2 + 4\left[\int_a^b \left|\frac{du}{dt} - i\,Au\right|^2 dt\right]^2.$$

Questa conduce conseguentemente ad un migliore risultato sulla convessità.

Dim. del Teorema 3.6. Noi possiamo assumere che $a = 0$, e $b = 1$, l'ultima si realizza sostituendo t con t/b e ciò non cambia le condizioni del teorema. Per semplificare la discussione ancora di più noi considereremo solo il caso $\alpha = \frac{\pi}{2}$. Estendiamo u nella striscia $\Sigma: 0 \leq \mathrm{Re}\ t \leq 1$ ponendo

$$(3.13) \qquad u(t) = u(s + ir) = T(r)u(s).$$

Applicando l'operatore di Cauchy-Riemann $\frac{\partial}{\partial \bar{t}} = \frac{1}{2}(\frac{\partial}{\partial s} + i\frac{\partial}{\partial r})$

troviamo

$$(3.14)\qquad \frac{\partial u(t)}{\partial \bar{t}} = \frac{i}{2}\, T(r)\, f(s).$$

Sui lati della striscia abbiamo $|u(t)| \le K(\,|u(o)| + |u(1)|\,)$ a causa della (3.7).

Poniamoci ora in condizioni di usare il "teorema delle tre rette" di Hadamard. A questo scopo sottraiamo da u una soluzione particolare della (3.14) in Σ. La differenza è allora olomorfa ed il teorema è applicabile. Come particolare soluzione scegliamo ($z = x+iy$, e $u_{\bar{t}}$ indica $\frac{\partial u}{\partial \bar{t}}$)

$$w(t) = -\frac{1}{\pi} \iint_{\Sigma} u_{\bar{t}}(z) \left(\frac{1}{z-t} + \frac{1}{\bar{z}+1+t} \right) dxdy .$$

Il nucleo dell'integrale è una soluzione fondamentale per l'operatore di Cauchy-Riemann e si verifica subito che l'integrale è assolutamente convergente e che

$$w_{\bar{t}}(t) = u_{\bar{t}}(t) .$$

Allora $h(t) = u(t) - w(t)$ è una funzione olomorfa in Σ.

Prima di procedere oltre osserviamo che vale la seguente disuguaglianza, facile da verificarsi, con qualche costante $C \ge 1$:

$$\int_{-\infty}^{\infty} \left| \frac{1}{z-t} + \frac{1}{\bar{z}+1+t} \right| dy < C \left(1 + \log \frac{1}{|x-s|}\right) \qquad \text{se } x \ne s.$$

Segue di qui e dalla (3.14), avendo posto

$$(3.15) \qquad \max_{0 \leq s \leq 1} \int_0^1 |f(x)| \, (1 + \log \frac{1}{|x-s|})dx = F,$$

che

$$(3.16) \qquad |w(t)| \leq \frac{C}{2} K F.$$

Consideriamo ora la funzione olomorfa h. Dalla (3.16) e dalla limitazione trovata per $u(t)$ sui lati di Σ, vediamo che sugli stessi lati

$$|h(t)| \leq K(|u(o)| + |u(1)| + \frac{C}{2} F).$$

Per il "teorema delle tre rette" la stessa disuguaglianza vale nell'interno di Σ, e segue che per t reale

$$|u(t)| \leq |h(t)| + |w(t)| \leq K (|u(o)| + |u(1)| + C F)$$

che è proprio la (3.12) cercata. Q.E.D.

Possiamo ora facilmente dedurre un risultato di convessità per le soluzioni della (3.2).

$$(3.17) \qquad |Lu| \leq \varphi(t) |u| .$$

Applicando la (3.12) abbiamo

$$\max_{a \leq t \leq b} |u(t)| \leq C K (|u(a)| + |u(b)| + B \max_{a \leq t \leq b} |u(t)|)$$

dove

$$B = \sup_{a \leq t \leq b} \int_a^b \varphi(x)\left(1+\log \frac{b-a}{|x-t|}\right) dx .$$

Se l'intervallo (a, b) è abbastanza piccolo in modo che $B\,K\,C \leq 1/2$ allora si ha

$$\max_{a \leq t \leq b} |u(t)| \leq 2\,C\,K(|u(a)| + |u(b)|). \tag{3.18}$$

Ora per ogni arbitrario numero reale ρ poniamo

$$v(t) = e^{-i\lambda t}\, u(t), \qquad \lambda = \rho e^{-i\alpha} \;;$$

v soddisfa la disuguaglianza

$$\left|\frac{1}{i}\frac{dv}{dt} + (\lambda - A_\alpha)\, v\right| \leq \varphi(t)\,|v(t)| .$$

Chiaramente $A_\alpha - \lambda = e^{-i\alpha}(A_0 - \rho) = e^{-i\alpha}\widetilde{A}$, e $i\widetilde{A}$ è un generatore infinitesimale del gruppo $\widetilde{T}(t) = e^{-i\rho t}\, T(t)$ il quale soddisfa anche la condizione $|\widetilde{T}(t)| \leq K$. Quindi la disuguaglianza (3.18) è anche valida per la funzione $v(t)$. Allora ponendo $\sigma = \rho \operatorname{sen} \alpha$ abbiamo per ogni numero reale σ

$$|u(t)| \leq 2C_2 K\left[|u(a)|\, e^{\sigma(t-a)} + |u(b)|\, e^{\sigma(t-b)}\right].$$

Tenendo T fisso e scegliendo la costante σ in modo che i due termini nel secondo membro siano uguali, otteniamo il seguente risultato di convessità:

<u>Cor.</u>: <u>Sia</u> u <u>una soluzione della</u> (3.17). <u>Se l'intervallo</u> (a, b) <u>è così piccolo che</u>

$$\sup_{a \le t \le b} \int_a^b \varphi(x)\left(1+\log\frac{b-a}{|x-t|}\right)dx \le \frac{1}{2CK} ,$$

allora per $a < t < b$, si ha

$$(3.19) \qquad |u(t)| \le 4\,C\,K\,|u(a)|^{\frac{b-t}{b-a}} \cdot |u(b)|^{\frac{t-a}{b-a}}$$

Applicando questo risultato ripetutamente a intervalli consecutivi possiamo ottenere limitazioni inferiori per le soluzioni della (3.17) per $t>0$. Per esempio si può dimostrare il seguente

Teorema 3.7. Se $\varphi(t) = H(1+t)^{-k}$, $k, H \ge 0$; allora una soluzione u della (3.17) soddisfa, per $t \ge 1$, la disuguaglianza

$$|u(t)| \ge |u(o)|\,\beta^{t}\, e^{-\mu t} \qquad \text{se } K > 1$$

$$|u(t)| \ge |u(o)|\,\beta^{t}\, e^{-\mu(t+1)^2} \qquad \text{se } K = 0$$

dove μ è una costante fissa e β dipende dalla soluzione.

In pratica si potrebbe applicare tale risultato ad una equazione differenziale a derivate parziali in un cilindro tale che diventi iperbolica quando si sostituisca $\frac{\partial}{\partial t}$ con $e^{i\alpha}\frac{\partial}{\partial t}$, cioè a quelle equazioni per le quali il problema iniziale e con condizioni al contorno è ben posto sia per valori positivi che per valori negativi del tempo.

Capitolo 4

Stabilità all'infinito.

4.1. In questo capitolo noi presentiamo un risultato relativo al problema (iv); supponiamo che Y sia uno spazio di Hilbert. Nel caso di dimensioni finite il risultato risale a Dunkel [5]. Nelle applicazioni di questo risultato ad operatori differenziali in cilindri come negli esempi 2, 3, 4 si considerano operatori differenziali i cui coefficienti possono dipendere da t ma tali che le differenze di essi coefficienti dai loro valori limiti (dipendenti solo da x) sono limitate da una costante per t^{-k} per qualche $k > 0$. Allora si conclude che le soluzioni che sono di quadrato integrabile sull'asse positivo tendono a zero esponenzialmente. Il numero k dev'essere preso almeno uguale all'ordine massimo dei poli reali della risolvente $R(\lambda)$.

Noi consideriamo soluzioni della disuguaglianza

$$(4.1) \qquad |Lu| \leq \frac{c}{(1+t)^k} |u|$$

che sono di quadrato sommabile per $t>0$, e assumeremo che $R(\lambda)$ sia regolare sull'intero asse reale eccettuato al più per un numero finito di poli reali $\lambda_1, \ldots\ldots, \lambda_m$, e che $|R(\lambda)| = 0(1)$ quando $|\lambda| \to \infty$ sull'asse reale.

Teorema 4.1. Se ciascuno dei poli $\lambda_1, \ldots\ldots, \lambda_m$ è di ordine $\leq h$ e se c è sufficientemente piccolo allora esistono due costanti positive a, C tali che per ogni soluzione della (4.1) di quadrato integrabile si ha:

$$(4.2) \qquad \int_0^\infty |e^{at} u|^2 \, dt < C \int_0^1 |u|^2 \, dt \quad .$$

Noi abbiamo visto nella sezione 1 che l'ipotesi su c non può essere eliminata.

La dimostrazione del teorema è basata sulla seguente disuguaglianza a priori.

Lemma 4.1. Supponiamo che $R(\lambda)$ soddisfi le condizioni precedenti. Sia $v(t)$ una funzione tale che $v(o) = o$ e $|v(t)|$ e $|(1+t^k)v(t)|$ siano di quadrato sommabile sull'asse positivo. Allora si ha

$$(4.3) \qquad \int_0^\infty |v|^2 dt < C_1 \int_0^\infty |(1+t^k) Lv|^2 dt$$

dove la costante C_1 dipende solo dall'operatore A.

Assumiamo dapprima il lemma e vediamo come da esso si deduca il teorema. Il teorema segue con un ben noto ragionamento dalla disuguaglianza :

$$(4.4) \qquad \int_T^\infty |u|^2 dt \leq C_2 \int_{T-1}^T |u(t)|^2 dt$$

dove $T > 1$; la costante C_2 è indipendente da T. Nel seguito $C_3, C_4, \dots$ sono costanti indipendenti da u e da T. Per dimostrare la (4.4.) consideriamo una funzione crescente indefinitamente derivabile $\zeta(t) \geq 0$ la quale sia uguale a zero per $t < 0$ ed uguale a uno per $t \geq 1$, e poniamo

$$v(t) = \zeta(t)\, u(t + T - 1) \qquad t \geq 0,$$

con $v(t) = 0$ per $t < 0$. Dal lemma abbiamo

$$\int_T^\infty |u|^2 dt \le \int_0^\infty |v|^2 dt \le C_1 \int_0^\infty |(1+t^k)\, Lv|^2 dt$$

$$\le C_1 \int_1^\infty |(1+t^k)\, Lu(t+T-1)|^2 dt +$$

$$+ C_3 \int_0^1 (1+t^k)^2 \left\{ |Lu(t+T-1)|^2 + |u(t+T-1)|^2 \right\} dt$$

con qualche costante C_3 dipendente da ζ ; quindi

$$\int_T^\infty |u|^2 dt \le C_1 \int_T^\infty (1+t^k)^2 |Lu(t)|^2 dt + C_4 \int_{T-1}^T (|Lu|^2 + |u|^2) dt.$$

Usando la (4.1) abbiamo

$$\int_T^\infty |u|^2 dt \le C_1 c^2 \int_T^\infty |u(t)|^2 dt + C_5 \int_{T-1}^T |u(t)|^2 dt$$

la quale dà la (4.4) se $C_1 c^2 \le \frac{1}{2}$.

4.2. Dimostrazione del Lemma 4.1 : $c_1, c_2, \ldots\ldots$ indicheranno costanti dipendenti soltanto da A . Se $Lv = f$ noi abbiamo come al solito per la trasformata di Fourier $(\lambda - A)\hat{v} = \hat{f}$, o

$$\hat{v}(\lambda) = R(\lambda)\, \hat{f}(\lambda)$$

dovunque $R(\lambda)$ sia regolare. Siano i poli $\lambda_1 < \lambda_2 < \ldots\ldots < \lambda_m$; decomponiamo $R(\lambda)$ in una somma finita per mezzo di una partizione finita dell'unità data da $(m+2)$ funzioni non negative di C^∞ : $\sigma_j(\lambda)$, $\sum_0^{m+1} \sigma_j(\lambda) \equiv 1$ le quali abbiano la proprietà che λ_i non appartiene al supporto delle σ_j eccettuato σ_i. I supporti di $\sigma_1, \ldots\ldots, \sigma_m$ sono limitati mentre i supporti di σ_o e σ_{m+1} si estendono a $-\infty$ ed a $+\infty$ rispettivamente. Poniamo

$$R_j(\lambda) = \sigma_j(\lambda) R(\lambda) \qquad\qquad w_j(\lambda) = R_j(\lambda)\hat{f}(\lambda) \quad ,$$

per modo che $\hat{v} = \sum w_j(\lambda)$. In un intorno di λ_j : $\hat{v} = w_j$ così che $|w_j|$ è di quadrato integrabile. Poichè $w_j(\lambda)$, $j = 1, \ldots, m$, ha supporto compatto essa è la trasformata di Fourier di una funzione analitica $v_j(t)$ che è in L_2. Noi dimostreremo per ogni j le disuguaglianze

$$(4.3') \qquad \int_0^\infty |v_j|^2 dt \leq \text{costante} \int_0^\infty |(1+t^k) f(t)|^2 dt .$$

Nel punto λ_j, $R(\lambda)$ ha un polo di ordine $\leq k$ e dunque può essere sviluppata in un intervallo contenente il supporto di σ_j nel modo seguente:

$$R(\lambda) = \sum_{r=1}^{k} (\lambda - \lambda_j)^{-r} P_r + P_o(\lambda)$$

dove P_r, $r=1, \ldots\ldots, k$ sono operatori limitati fissati e $P_o(\lambda)$ è un operatore olomorfo nel rettangolo:

$\{\text{Re}\,\lambda$ in un intervallo aperto contenente il supporto di σ_j, e $|\text{Im}\,\lambda| < \varepsilon\}$.

Per la formula integrale di Cauchy abbiamo che le derivate

$(\frac{d}{d\lambda})^r P_o(\lambda)$, con $r \leq k$, sono limitate in norma nel supporto di σ_j. Allora abbiamo

$$(\lambda-\lambda_j)^k w_j(\lambda) = \sum_1^k (\lambda-\lambda_j)^{k-r} \sigma_j(\lambda) P_r \hat{f}(\lambda) + \sigma_j(\lambda)(\lambda-\lambda_j)^k P_o(\lambda)\hat{f}(\lambda).$$

Differenziando questa uguaglianza e tenendo conto del fatto che le derivate di $P_o(\lambda)$ e di $\sigma_j(\lambda)$ sono limitate abbiamo

$$|(\frac{d}{d\lambda})^n(\lambda-\lambda_j)^k w_j(\lambda)| \leq c_1 \sum_{r=o}^{n} |(\frac{d}{d\lambda})^r \hat{f}(\lambda)|, \qquad \text{per } n \leq k.$$

Poichè la trasformata di Fourier inversa di $(\frac{d}{d\lambda})^n(\lambda-\lambda_j)^k w_j(\lambda)$ è

$$(\frac{t}{i})^n(D_t-\lambda_j)^k v_j(t) = (\frac{t}{i})^n e^{i\lambda_j t} D_t^k (e^{-i\lambda_j t} v_j),$$

segue dal teorema di Parseval che

$$\int_{-\infty}^{\infty} |(1+t^k) D_t^k (e^{-i\lambda_j t} v_j)|^2 dt \leq c_2 \int_0^{\infty} (1+t^k)^2 |f(t)|^2 dt \quad .$$

Applichiamo ora una ben nota disuguaglianza di Hardy (cfr. [6] teorema 330) secondo la quale per funzioni scalari $a(t)$ si ha

$$\int_0^{\infty} |a(t)|^2 dt \leq \text{costante} \int_0^{\infty} |t^k D_t^k a|^2 dt$$

(ciò si prova facilmente con ripetute integrazioni per parti).

Usando le due ultime disuguaglianze abbiamo

$$\int_0^\infty |v_j|^2 dt = \int_0^\infty |e^{-i\lambda_j t} v_j|^2 dt \leq c_3 \int_0^\infty (1+t^k)^2 |f(t)|^2 dt \quad ,$$

cioè la disuguaglianza (4.3') cercata.

Finalmente consideriamo le funzioni $w_j = R_j(\lambda)\hat{f}(\lambda)$ per $j=0$ e $m+1$. Siccome $|R(\lambda)|$ è uniformemente limitata lontano dai poli noi deduciamo come prima che w_0 e w_{m+1} sono trasformate di Fourier di funzioni di quadrato integrabile $v_0(t)$, $v_{m+1}(t)$ le quali soddisfano la disuguaglianza

$$\int_0^\infty |v_j|^2 dt \leq c_4 \int_0^\infty |f(t)|^2 dt \quad , \qquad j=0 \text{ e } (m+1)$$

quindi in particolare anche la (4.3'). Q.E.D.

Osserviamo che se $R(\lambda)$ soddisfa inoltre la ulteriore condizione $\left|\lambda \frac{dR}{d\lambda}\right| = 0(1)$ per $|\lambda| \to \infty$ sull'asse reale, nel teorema e nel lemma precedenti si può sostituire la norma in L_2 con quella in L_p, per $1 < p < \infty$.

B I B L I O G R A F I A

[1] S. Agmon — On the eigenfunctions and on the eigenvalues of general elliptic boundary value problems, Comm. Pure Appl. Math. 15 (1962), pp. 119-147.

[2] S. Agmon, L. Nirenberg - Properties of solutions of ordinary differential equations in Banach space. Comm. Pure Appl. Math. 16 (1963) pp. 121-239.

[3] P. Cohen, M. Lees - Asymptotic decay of solutions of differential inequalities, Pacific J. Math. 11 (1961) pp. 1235-1249.

[4] N. Dunford, J. T. Schwartz - Linear Operators I, Interscience Publications, New York 1958.

[5] O. Dunkel — Regular singular points of a system of homogeneous linear differential equations of the first order, Proc. Amer. Acad. Sci. 38 (1912-1913) pp. 341-370.

[6] G. H. Hardy, J. E. Littlewood, G. Polya - Inequalities. Cambridge Univ. Press, London 1951.

[7] S. G. Krein, O. I. Prozorovskaya - Analytic semigroups and incorrect problems for evolutionary equations, Doklady Akad. Nauk. SSSR, N.S. 133 (1960) pp. 277-280. English translation, Soviet Mathematics Amer. Math. Soc. 1 (1960) pp. 841-844.

[8] P. D. Lax — A stability theorem for solutions of abstract differential equations and its application to the study of local behavior of solutions of elliptic equations. Comm. Pure Appl. Math. 9 (1956) pp. 747-766.

[9] __________ A Phragmén-Lindelöf principle in harmonic analysis with application to the separation of variables in the theory of elliptic equations. Lecture Series of Sympo-

sium on Partial Differential equations, Berkeley, Calif., summer 1955. Univ. of Kansas 1957.

[10] ______________ A Phragmén-Lindelöf theorem in harmonic analysis and its application to some questions in the theory of elliptic equations. Comm. Pure Appl. Math. 10 (1957) pp.361-389.

[11] Yu.I. Lyubič. Conditions for the uniqueness of the solution to Cauchy's abstract problem. Doklady Akad. Nauk. SSSR N.S. 130 (1960) pp. 969-972. English translation, Soviet Mathematics Amer. Math. Soc. 1 (1960) pp. 110-113.

[12] A. Plis A smooth linear elliptic differential equation without any solution in a sphere. Comm. Pure Appl. Math. 14 (1961) pp. 599-617.

[13] E. C. Titchmarsh The theory of functions. The Clarendon Press, Oxford, 1939.

CENTRO INTERNAZIONALE MATEMATICO ESTIVO

(C.I.M.E.)

R.S. PHILLIPS

SEMI-GROUPS OF CONTRACTION OPERATORS

Roma - Istituto Matematico dell'Università

SEMI-GROUPS OF CONTRACTION OPERATORS

by

R.S. PHILLIPS

Resume :

1 - Introduction : The initial-value problem as motivation for the theory of semi-groups of operators.

2 - Strongly-continuous semi-groups of operators : a short course in the theory of semi-groups of operators.

3 - Dissipative operators, the class of generators of semi-groups of contraction operators. A Cayley-transform theory for dissipative operators.

4 - The double extension construction for pairs of dissipative operators contained in each other's adjoints.

5 - Boundary space theory relative to a given null subspace.

6 - Operator theory : the application of boundary space theory to operators and in particular to symmetric system of differential operators.

R. S. Phillips

Semi-Groups of Contraction Operators

1. Introduction.

The theory of semi-groups of linear operators had its origin in Stone's theorem on groups of unitary operators acting in a Hilbert space (1932). Stone's theorem was motivated by the time dependent solution of Schroedinger's wave equation for quantum mechanics. The study of semi-groups of operators, rather than groups of operators, was undertaken by Hille in 1936 who became interested in the semi-group properties of certain classical singular integrals. However it was not until 1948 that the applicability of the theory was fully appreciated. At that time K. Yosida applied semi-group methods to the diffusion equation. In the hands of Feller, Hille, Kendall, Reuter, and Yosida, the theory became an integral part of the theory of probability. It has also been applied with profit to the Cauchy problem for the wave equation; here the contributions of Friedrichs, Lax, and Phillips should be mentioned.

We shall use the initial value problems of mathematical physics to motivate the theory of semi-groups. A suitably abstract formulation can be obtained by the following considerations.

It will be recalled that Hadamard called an initial value problem <u>well set</u> if

i) There is a unique solution to the problem for some given class of initial data,

ii) The solution varies continuously with the initial data. These two requirements are eminently reasonable on physical grounds.

The existence and uniqueness of the solution being an affirmation of the <u>principle of scientific determinism</u>, whereas the continuous dependence is an expression of the stability of the solution.

Stability is essential in view of the impossibility of knowing the precise initial state and the consequent desirability of approximating the true solution with approximate initial data.

If D_t denotes the set of initial data for the problem at time $t \geq 0$, then the state of the system with initial data y in D_{t_0} is determined for all $t > t_0$. Denoting this state by $S(t, t_0)y$, it is clear that the solution will be continuable only if we require that $S(t, t_0)y$ in turn belong to D_t.

We can now compute $S(t_0 + \tau_1 + \tau_2, t_0)y$ either directly $(\tau_1, \tau_2 > 0)$ or indirectly, using $S(t_0 + \tau_1, t_0)y$ as initial data and obtaining the solution at a time τ_2 later as

$$S(t_0 + \tau_1 + \tau_2, t_0 + \tau_1) \; S(t_0 + \tau_1, t_0)y.$$

The uniqueness condition (i) then implies the semi-group property

$$S(t_2, t_1)\, S(t_1, t_0) = S(t_2, t_0), \qquad t_0 < t_1 < t_2.$$

As was pointed out by Hadamard, this semi-group property is reflected in certain addition theorems.

Stability requires that $\lim_n S(t, t_0)y_n = S(t, t_0)y$ whenever $y_n \longrightarrow y$, assuming of course that all of the data involved belong to D_{t_0}. In other words $S(t, t_0)$ is a continuous operator on D_{t_0}. Moreover to say that y is the initial value for a solution means that $\lim_{t \to t_0} S(t, t_0)y = y$. In the usual terminology this means that operators $S(t, t_0)$ converge strongly to the identity as $t \longrightarrow t_{0+}$.

In addition to the above mentioned Hadamard conditions, we now assume that the problem is basically time-invariant. Physically this means that the underlying mechanism does not depend on time or, equivalently, that the corresponding differential equation including the boundary conditions are time invariant. In terms of the above notation this amounts to

iii) D_t is independent of t ,

iv). $S(t_2, t_1)$ depends only on $t_2 - t_1$.

We shall call the common set of initial data D and set $S(\tau) = S(t + \tau, t)$.

The semi-group property then becomes

$$S(t_1 + t_2) = S(t_1)S(t_2) , \qquad t_1, t_2 > 0. \tag{1}$$

In some physical problems the initial data determines the entire past as well as the future of the system. For such mechanisms the restriction $t_1, t_2 > 0$ need not be imposed and (1) will hold for all real t_1, t_2. The resulting family of operators defines a <u>group</u> of operators.

We shall also assume that

v). $S(t)$ is linear.

Condition (v) will satisfied if the associated differential operator is linear and the boundary conditions determining D are homogeneous.

Thus existence, uniqueness, stability, time-invariance, and linearity constitute our principal assumptions. It is clear that a large class of initial value problems from mathematical physics satisfy these requirements and consequently have solutions which can be described by strongly continuous semi-groups of linear operators.

Implicit in the above discussion is a topology on D. We shall henceforth assume that D lies in a Banach space X and without loss of generality we may assume that D is dense in X (since otherwise we can take X to be the closure of D). If for fixed t , the operator S(t) is linear and continuous on D , then by the usual argument it can be shown that S(t) has a unique linear bounded extension on all of X . We denote the so-extended operator again by S(t) and it is readily verified that the resulting operators again have the semi-group property .

The condition

$$\lim_{t \to 0} S(t)y = y, \qquad y \in D,$$

is not sufficient for the development of a simple theory. There are many ways to supplement this hypothesis, the simplest being

(vi) $$\lim_{t \to 0} S(t)y = y, \qquad y \in X.$$

Equivalently we could assume taht S(t) is bounded in norme near the origin (by the uniform boundedness theorem).

This condition is satisfied by pratically all of the applications of the theory.

2. Strongly continuous semi-groups of operators.

We shall now derive some of the elementary properties of strongly continuous semi-groups of operators. These are one-parameter families of bounded operators subject to
the following conditions :

i). $S(t_1 + t_2) = S(t_1)S(t_2),\qquad t_1, t_2 \geqslant 0,$

$S(0) = I;$

ii). $\lim_{t \to 0+} S(t)y = y, \qquad y \in X.$

It follows from the uniform boundedness theorem that

$$|S(t)| \leqslant M \qquad 0 \leqslant t \leqslant \delta$$

for some positive constants M and δ. Since any positive t can be written as $t = n\delta + \tau$, $0 \leqslant \tau < \delta$, we see that

$$|S(t)| \leqslant |S(\delta)|^n \, |S(\tau)| \leqslant M^{n+1} \leqslant M\, e^{\omega t}$$

where $\omega = (\log M)/\delta$. Actually more is true.

Lemma 1. $\omega_0 \equiv \inf_{t>0} (\log |S(t)|)/t = \lim_{t \to \infty} (\log |S(t)|)/t.$

Proof. Setting $f(t) = \log|S(t)|$, we see that

$$f(t_1 + t_2) = \log |S(t_1 + t_2)| \leqslant \log |S(t_1)| \, |S(t_2)| = f(t_1) + f(t_2),$$

so that $f(t)$ is subadditive. Moreover, as we have just remarked $f(t) \leqslant \log M + \omega t$. Now $\beta = \inf_{t>0} f(t)/t$ is either finite or $-\infty$. In the former case, given $\varepsilon > 0$ there is an $\eta > 0$ such that $\beta \leqslant f(\eta)/\eta \leqslant \beta + \varepsilon$. Hence for $(n-1)\eta < t < n\eta$, we have

$$f(t)/t \leqslant \left[(n-1)f(\eta) + f(t-(n-1)\eta)\right]/t \leqslant f(\eta)/\eta + \left[\log M + \omega\eta\right]/t.$$

It follows that $\limsup_{t\to\infty} f(t)/t \leqslant \beta + \varepsilon$, and hence that $\lim_{t\to\infty} f(t)/t = \beta$. The case $\beta = -\infty$ is treated in a similar way.

The parameter ω_0 plays an important role in the theory of semi-groups of operators. It is called the <u>type</u> of $[S(t)]$.

For one thing $\exp(\omega_0 t)$ is the spectral radius of the operator $S(t)$; in fact

$$\left|S(t)^n\right|^{1/n} = \exp\left[\frac{t}{nt}\log\left|S(nt)\right|\right] \to \exp\ \omega_0 t.$$

Also it is clear that for each $\omega_1 > \omega_0$ there exists an $M > 0$ such that

$$\left|S(t)\right| \leqslant M\exp(\omega_1 t).$$

Lemma 2. The semi-group $[S(t)\ ;\ t \geqslant 0]$ is strongly continuous.

Proof. For arbitrary $y \in X$ and $t_2 > t_1 \geqslant 0$, we have

$$\left|S(t_2)y - S(t_1)y\right| \leqslant \left|S(t_1)\right|\left|S(t_2-t_1)y - y\right|$$

$$\leqslant M\exp(\omega t_1)\left|S(t_2-t_1)y - y\right|$$

and the right member tends to zero as $t_2 - t_1 \to 0$.

The initial-value problem is usually presented in the form of a differential equation of the type $d\,S(t)y/dt$ equal to a spatial operator where y belongs to the set of initial data D. In the process of abstracting the defining conditions (i) and (ii) we lost track of D. On the other hand we cannot expect an arbitrary $y \in X$ to serve as initial data for the class of problems we wish to consider. Fortunately it is possible to retrieve D from the semi-group of operators. This is accomplished by means of the infinitesimal generator A , defined as follows :

Setting

$$A_\eta = \frac{S(\eta) - I}{\eta} \qquad \eta > 0,$$

we define

$$Ay = \lim_{\eta \to 0^+} A_\eta y$$

whenever this limit exists. The set of elements y for which the limit exists is the domain of A , denoted by D_A . It is clear that D_A is a linear subspace and that A is a linear operator. It turns out that D_A corresponds to the above mentioned set of initial data D . We now show that D_A is dense in X and that A is a closed operator.

Lemma 3. D_A is dense in X .

Proof. We set

$$y_\alpha = \int_0^\alpha S(t)y \, dt.$$

It is clear that $\alpha^{-1} y_\alpha \to \alpha$ as $\alpha \to 0^+$ and hence that the elements $\{y_\alpha\}$ are dense in X . On the other hand

$$A_\eta y_\alpha = \eta^{-1} \int_0^\alpha \left[S(t+\eta)y - S(t)y\right] dt$$

$$= \eta^{-1} \int_\alpha^{\alpha+\eta} S(t)y \, dt - \eta^{-1} \int_0^\eta S(t)y \, dt$$

so that $\lim_{\eta \to 0+} A_\eta y_\alpha = S(\alpha)y - y$. Hence the y_α belong to D_A.

Lemma 4. If $y \in D_A$, then so does $S(t)y$ and

(2) $$d\, S(t)y/dt = AS(t)y = S(t)Ay, \qquad t > 0.$$

Proof. For $\eta > 0$, we have

$$A_\eta S(t)y = \eta^{-1}\left[S(t+\eta)y - S(t)y\right] = S(t)A_\eta y.$$

Assuming $y \in D_A$, we see that $S(t)y \in D_A$ and that

$$AS(t)y = d^+ S(t)y/dt = S(t)Ay.$$

Moreover $\eta^{-1}\left[S(t)y - S(t-\eta)y\right] = S(t-\eta)A_\eta y$, and since $S(t)$ is strongly continuous, the limit as $\eta \to 0+$ exists and

$$d^- S(t)y/dt = S(t)Ay.$$

Corollary. For $y \in D_A$,

(3) $$S(t)y - y = \int_0^t S(\tau)Ay\, d\tau = \int_0^t AS(\tau)y\, d\tau = A\int_0^t S(\tau)y\, d\tau;$$

Proof. The first two equalities follow from integrating the relation (2) and the third from the proof. of Lemma 3.

Lemma 5. A is a closed linear operator.

Proof. Suppose the sequence $\{y_n\} \subset D_A$, that $y_n \to y$ and that $Ay_n \to z$. Then $S(\tau)Ay_n \to S(\tau)z$ uniformly on $[0,\eta]$ and hence by (3)

$$A_\eta y = \eta^{-1}\left[S(\eta)y - y\right] = \eta^{-1}\int_0^\eta S(\tau)z\, d\tau \to z \qquad \text{as} \quad \eta \to 0+.$$

As a consequence $y \in D_A$ and $Ay = z$, showing that A is closed.

We have now returned to our starting point, namely, the initial value problem. Setting $D = D_A$, we have shown that D is dense in X, that $S(t)y$ provides a solution to the initial value problem

$$dS(t)y/dt = AS(t)y$$

$$S(0)y = y$$

for y in D which possess all of the desidered features. Moreover this gives the only solution to this problem. For if $y(t)$ is a strongly continuously differentiable function on $(0, \infty)$ to X such that

$$(4) \qquad \begin{aligned} &d\, y(t)/dt = Ay(t), && t > 0, \\ &\lim_{t \to 0+} y(t) = y_0 , \end{aligned}$$

then $y(t) = S(t)y_0$. In fact for $0 < \tau < t$, the function $S(t - \tau)y(\tau)$ is strongly continuously differentiable in τ with derivative

$$d\, S(t - \tau)y(\tau)/d\tau = S(t-\tau)dy(\tau)/d\tau - S(t - \tau)Ay(\tau) = 0.$$

The desired result follows on integrating from 0 to t.

The classical procedure in solving the initial-value problem (4) is to take the Laplace transform. Heuristically $y(t) = S(t)y = [\exp(tA)]\, y$ and

$$\int_0^\infty \exp(-\lambda t)S(t)\, dt = \int_0^\infty \exp(-\lambda I - A)t\, dt = (\lambda I - A)^{-1} = R_\lambda(A).$$

We now derive this result rigorously.

Theorem 1. Let $[S(t)]$ be a semi-group of type ω_0 and infinitesimal generator A. Then for all λ with $re\,\lambda > \omega_0$,

$$R_\lambda(A)y = \int_0^\infty \exp(-\lambda t)\, S(t)y\, dt\,, \qquad y \in X.$$

Proof. For a λ with re $\lambda > \omega_1 > \omega_0$, $\lambda = \sigma + i\nu$,

$$|S(t)y| \leqslant M\,|y|\, \exp(-[\sigma - \omega_1]\, t).$$

As a consequence, $\exp(-\lambda t)\, S(t)y$ is integrable and

$$Ry = \int_0^\infty \exp(-\lambda t)\, S(t)y\, dt$$

defines a linear bounded operator with norm $|R| \leqslant M/(\sigma - \omega_1)$.

Next we show that $Ry \in D_A$. In fact for $\eta > 0$

$$A_\eta R_y = \eta^{-1} \int_0^\infty \exp(-\lambda t) \left[S(t+\eta)y - S(t)y\right] dt$$

$$= \frac{e^{\lambda\eta}}{\eta} \int_0^\infty e^{-\lambda(t+\eta)}\, S(t+\eta)y\, dt - \frac{1}{\eta}\int_0^\infty e^{-\lambda t}\, S(t)y\, dt$$

$$= \frac{e^{\lambda\eta} - 1}{\eta} \int_0^\infty e^{-\lambda t}\, S(t)y\, dt - \frac{e^{\lambda\eta}}{\eta} \int_0^\eta e^{-\lambda t}\, S(t)y\, dt\,.$$

As $\eta \to 0+$,

$$\frac{e^{\lambda\eta} - 1}{\eta} \to \lambda \quad \text{and} \quad \frac{1}{\eta}\int_0^\eta e^{-\lambda t}\, S(t)y\, dt \to y.$$

Conseguently $Ry \in D_A$ and

(5) $$AR_y = \lambda R_y - y$$

On the other hand, for $y \in D_A$, $A \exp(-\lambda t) S(t)y = \exp(-\lambda t) S(t)Ay$ is strongly continuous and majorized by $M \|Ay\| \exp(-[\sigma - \omega_1] t)$. Thus both $\exp(-\lambda t) S(t)y$ and $A \exp(-\lambda t) S(t)y$ are integrable and since A is closed we may conclude that

$$A \int_0^\infty e^{-\lambda t} S(t)y \, dt = \int_0^\infty e^{-\lambda t} AS(t)y \, dt = \int_0^\infty e^{-\lambda t} S(t)Ay \, dt.$$

In other words $ARy = RAy$. Combining this with (5), we see that

$$(\lambda I - A)Ry = y, \qquad y \in X,$$
$$R(\lambda I - A)y = y, \qquad y \in D_A,$$

from which it follows that $R = R_\lambda(A)$, the resolvent of A.

The previous theorem shows that $\mathrm{re}\,[\sigma(A)] \leqslant \omega_0$; here we use $\sigma(A)$ to denote the spectrum of A. In particular, then, the resolvent set of any infinitesimal generator contains a right half plane.

The problem of when a closed linear operator is the infinitesimal generator of a semi-group of operators is basic in the applications of the theory. As we have seen, the resolvent $R_\lambda(A)$ is the Laplace transform of $S(t)$ when A is its generator. It is natural to ask for conditions on the resolvent of an operator which suffice to make this operator a Laplace transform.

This problem has its classical counterpart in the numerical case and the criteria which have been developed for generators have much the same form as the classical criteria. However the method of proof is quite different since the usual compactness arguments are not available; this deficiency is more than balanced by the special properties peculiar to the resolvent of an operator.

The first and in many ways the most useful generation theorem was obtained independently by E . Hille and K. Yosida in 1948. We shall derive their result as a corollary of.

Theorem 2. A necessary and sufficient condition for a closed linear operator U with dense domain to generate a semi-group $[S(t)]$ is that there exist real constants $M > 0$ and ω such that

$$\left| \left[R_\lambda (U) \right]^n \right| \leqslant M/(\lambda - \omega)^n \tag{6}$$

for all real $\lambda > \omega$ and $n = 1, 2, \ldots$. In this case

$$|S(t)| \leqslant M\, e^{\omega t}, \qquad t > 0. \tag{7}$$

Proof. Suppose first that $S(t)$ is a semigroup with generator A, bounded as in (7). According to the previous theorem,

$$R_\lambda (A)y = \int_0^\infty e^{-\lambda t}\, S(t)y\, dt, \qquad \lambda > \omega.$$

It is easy to see that one can interchange the order of integration and differentiation and so abtain

$$R_\lambda^{(n)} (A)y = (-1)^n \int_0^\infty t^n e^{-\lambda t}\, S(t)y \quad dt.$$

Consequently

$$\left| R_\lambda^{(n)} (A)y \right| \leqslant \int_0^\infty t^n e^{-\lambda t} M\, e^{\omega t}\, dt = n!\; M/(\lambda - \omega)^{n+1}.$$

On the other hand it is known for resolvent operators that

$$R_\lambda^{(n)}(A) = (-1)^n\, n!\, \left[R_\lambda(A)\right]^{n+1}$$

and combining this with the previous inequality gives (6).

The converse argument which we present is modelled after K. Yosida's proof of the Hille-Yosida theorem. We shall divide the proof into a number of steps.

a). Setting $B_\lambda = \lambda^2 R_\lambda(U) - \lambda I$, we show that

$$\lim_{\lambda \to \infty} B_\lambda y = Uy, \qquad y \in D_U.$$

Now $B_\lambda y = \lambda(\lambda R_\lambda(U)y - y) = \lambda R_\lambda(U)Uy$ so that it suffices to show that $\lambda R_\lambda(U)x \to x$ for all $x \in X$. Again if $x \in D_U$ then

$$|\lambda R_\lambda(U)x - x| = |R_\lambda(U)Ux| \leq K|Ux|/\lambda \to 0$$

as $\lambda \to \infty$; here we have used the inequality (6) for the case $n = 1$. Approximating an arbitrary x in X by a sequence in D_U and using the fact that the operators $[\lambda R_\lambda(U)]$ are uniformly bounded in norm for λ sufficiently large, the result now follows by the double limit theorem.

b). For each $\beta > 1$ there is a Λ_β such that

$$|\exp(t B_\lambda)| \leq M \exp(\beta \omega t), \qquad \lambda > \Lambda_\beta.$$

In fact

$$e^{tB_\lambda} = e^{-\lambda t} e^{t\lambda^2 R_\lambda(U)} = e^{-\lambda t} \sum_{n=0}^{\infty} \frac{t^n \lambda^{2n}}{n!} \left[R_\lambda(U)\right]^n$$

and making use of (6) we get

$$|e^{tB_\lambda}| \leq e^{-\lambda t} \sum_{0}^{\infty} \frac{t^n \lambda^{2n}}{n!} \frac{M}{(\lambda - \omega)^n} = M \exp\left(t\,\omega \frac{\lambda}{\lambda - \omega}\right),$$

It sufficies to choose Λ_β so that $\lambda(\lambda-\omega)^{-1} \leqslant \beta$ for $\lambda > \Lambda_\beta$.

c). $\lim_{\lambda\to\infty} \exp(tB_\lambda)y$ exists for each $y \in X$ uniformly with respect to t in each finite subinterval of $[0,\infty]$. To prove this we define the auxiliary function.

$$V(\tau) = \exp((t-\tau)B_\lambda)\exp(\tau B_\mu)$$

Both factors are continuously differentiable with respect to τ in the uniform operator topology and

$$dV(\tau)/d\tau = \exp((t-\tau)B_\lambda)(B_\mu - B_\lambda)\exp(\tau B_\mu) = V(\tau)\,(B_\mu - B_\lambda)$$

since $R_\lambda(U)$ and $R_\mu(U)$ commute. Integrating from 0 to t we get

$$\exp(tB_\mu) - \exp(tB_\lambda) = \int_0^t V(\tau)\,(B_\mu - B_\lambda)\,d\tau .$$

If we now make use of the estimate obtained in (b) we have

$$\left|\exp(tB_\mu)y - \exp(tB_\lambda)y\right| \leqslant \int_0^t M^2 \exp((t-\tau)\beta\omega)\exp(\tau\beta\omega)\left|B_\mu y - B_\lambda y\right| d\tau$$

$$\leqslant M^2 \exp(\beta\omega t)\, t\left|B_\mu y - B_\lambda y\right|.$$

If $y \in D_U$, then according to step (a) this converges to zero uniformly in t in each finite subinterval of $[0,\infty)$. Finally an arbitrary y in X can be approximated by a sequence in D_U and using the fact that the operators $[\exp(tB_\lambda)]$ are uniformly bounded in norm for sufficiently large, the result follows by the double limit theorem.

d). Setting $S(t)y = \lim_{\lambda\to\infty} \exp(tB_\lambda)y$, we now show that

$[S(t); t \geq 0]$ is a semi-group of operators. It is obvious that the approximating operators, namely $[\exp(tB_\lambda)]$ are semi-groups. Hence

$$S(t_1 + t_2)y = \lim_{\lambda \to \infty} \exp(t_1 B_\lambda) \exp(t_2 B_\lambda)y = S(t_1)S(t_2)y$$

since for strong limits the limit of a product is equal to the product of the limits. Further since the limit is uniform in t on subintervals, it follows that $S(t)y$ is continuous in $t \geq 0$; in particular $S(t)y \to y$ as $t \to 0+$. Finally we note that the inequality (7) is an immediate consequence of (b).

e). U is the infinitesimal generator of $[S(t)]$. It is clear that

$$\exp(tB_\lambda)y - y = \int_0^t \exp(\tau B_\lambda)B_\lambda y \, d\tau .$$

For $y \in D_U$ we have $\lim_{\lambda \to \infty} B_\lambda y = Uy$ so that the integrand converges uniformly on $[0, t]$ and we get

$$S(t)y - y = \int_0^t S(\tau)Uy \, d\tau .$$

Consequently

$$A_\eta y = \eta^{-1} \int_0^\eta S(\tau)Uy \, d\tau \to Uy$$

as $\eta \to 0+$. Denoting the generator of $[S(t)]$ by A, it follows that $D_A \supset D_U$ and $Ay = Uy$ on D_U. On the other hand for $\lambda > \omega$, $R_\lambda(U)$ exists by hypothesis and $R_\lambda(A)$ exists by Theorem 1. Hence

$$(\lambda I - U)D_U = (\lambda I - A)D_A$$

and this shows that $D_U = D_A$ and therefore that $A = U$. This concludes the proof of Theorem 2.

An operator of norm less than or equal to one is called a contraction operator. Semi-groups of contraction operators will constitute the main theme of these lectures. For such semi-groups the generation theorem takes on a particularly simple form.

Corollary (Hille-Yosida) A necessary and sufficient condition for a closed linear operator U with dense domain to generate a semi-group of contraction operators is that

$$|R_\lambda(U)| \leqslant 1/\lambda \qquad \lambda > 0. \tag{8}$$

Proof. If $M = 1$, it is readily seen that (8) implies (6) with $\omega = 0$. The assertion now follows directly from Theorem 2.

An example will serve to illustrate this theorem. For $x = C_0(-\infty, \infty)$, the space of continuous complex-valued functions which tend to zero at infinity, consider the following initial-value problem :

$$\frac{\partial u}{\partial t} = \frac{\partial^2 u}{\partial x^2}, \qquad u(0, x) = f(x), \qquad -\infty < x < \infty$$

We show that this has a semi-group solution with generator

$$Uy = \frac{d^2 y}{dx^2}; \qquad D_U = \text{class of functions } y(x) \text{ with } y, \frac{dy}{dx} \text{ continuously differentiable and } y, \frac{d^2y}{dx^2} \text{ in } X.$$

It is clear that D_U is dense in X. A solution of

$$(\lambda I - U)y = \lambda y - \frac{d^2 y}{dx^2} = f$$

can be obtained by the method of variation of parameters in the form

$$y(x) = \frac{1}{2\sqrt{\lambda}} \int_{-\infty}^{\infty} e^{-\sqrt{\lambda}\,(x-\sigma)} f(\sigma)\,d\sigma .$$

One verifies that $y \in D_U$ and that $\lambda y - \frac{d^2y}{dx^2} = f$. It remains only to show that $|R_\lambda(U)| \leqslant \lambda^{-1}$ for $\lambda > 0$. This estimate can be obtained directly from the integral representation of R_λ.

3 - Dissipative operators.

The theory of dissipative operators is intimately connected with semi-groups of contraction operators. Such semi-groups appear as the solutions to initial-value problems where a basic quantity does not increase in time. For hyperbolic equations this quantity is the energy, for parabolic equations it is the mass density, and for Markov processes it is the probability density. The study of all of these phenomena can be subsumed under the theory of dissipative operators. Unfortunately, however, not too much is known about dissipative operators in a Banach space. Most of the results apply only to Hilbert spaces; the results in l_1 are meager but promising and those for l_p, $1 < p < \infty$, are very sketchy. We shall limit our discussion to dissipative operators on a Hilbert space and their applications to certain initial-value problems.

Suppose $[S(t)\,;\, t \geqslant 0]$ is a semi-group of contraction operators with generator A. Then for $y \in D_A$ we have

$$0 \geqslant \frac{|S(\eta)y|^2 - |y|^2}{\eta} = \left(\frac{S(\eta)y - y}{\eta},\ S(\eta)y\right) + \left(y,\ \frac{S(\eta)y - y}{\eta}\right)$$

$$\rightarrow (Ay, y) + (y, Ay)$$

as $\eta \to 0$. Geometrically $\operatorname{Re}(Ay, y) \leqslant 0$ means that $Ay + y$ lies on the origin side of the tangent plane to the sphere $[z; |z| = |y|]$ at the point y. This forces the path of $y(t)$ for $\frac{dy}{dt} = Ay$ to stay within this sphere.

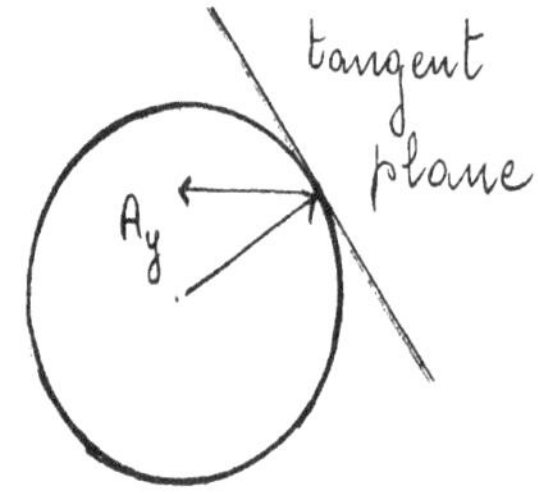

Definition. A linear operator L will be called <u>dissipative</u> if

$$(y, Ly) + (Ly, y) \leqslant 0 , \qquad y \in D_L ,$$

and <u>maximal dissipative</u> if it has no proper dissipative extension.

The salient result here is

Theorem 3. A necessary and sufficient condition for an operator L with dense domain to generate a semi-group of contraction operators is that L be maximal dissipative.

This can be considered as an existence theorem. Thus given a dissipative operator L_0 with dense domain, the theorem together with Zorn's lemma then asserts that there is a dissipative extension L of L_0 wich generates a semi-group of contraction operators. For example, let $H = l_2$

$$D(L_0) = \text{ultimately zero vectors } y = \{\eta_j\}$$

$$(L_0 y)_i = \sum a_{ij} \eta_j$$

where it is assumed that $\sum_i |a_{ij}|^2 < \infty$ for all j and for all finite sets of integers $\{n_i\}$ that the matrices

$$(a_{n_i n_j}) + (\bar{a}_{n_j n_i}) \leqslant \Theta.$$

In this case L_o is dissipative with dense domain and hence by the theorem there exists a dissipative extension which generates a semi-group of contraction operators.

Before proving Theorem 3., we first prove a basic lemma.

Lemma 6. If L is dissipative and $\lambda > 0$, then for

$$f = \lambda y - Ly$$

we have

$$\lambda|y| \leqslant |f| . \tag{9}$$

Proof. Clearly

$$2\lambda(y, y) \leqslant 2\lambda(y, y) - \left[(Ly, y) + (y, Ly)\right] = (f, y) + (y, f) \leqslant 2|y||f|$$

from which (9) follows.

Remark 1. One consequence of (9) is that $(\lambda I - L)^{-1}$ is a bounded operator on the range of $(\lambda I - L)$ which is a closed subspace if and only if the operator $(\lambda I - L)^{-1}$ is closed. Thus $(\lambda I - L)$ is a closed operator if and only if the range of $(\lambda I - L)$ is a closed subspace.

Remark 2. The map : $y \rightarrow f = y - Ly$ is one-to-one so that if the range of $I - L$ is all of H, then L is necessariely maximal dissipative. As we shall see the converse is also true for dissipative operators with dense domains.

Next we develop a Cayley transform theory which has Theorem 3 as one of its consequences. It will also give us some insight into the construction of the dissipative extension of a given dissipative operator. We define

$$J = (I + L)(I - L)^{-1}$$
$$D_J = \text{range of } I - L ,$$

and show that J is a contraction operator. In fact for $u \in D_J$

$$(10) \qquad u = y - Ly$$
$$Ju = y + Ly$$

for some $y \in D_L$. Hence

$$(Ju, Ju) = (y, y) + (Ly, Ly) + \left[(Ly, y) + (y, Ly)\right]$$
$$(u, u) = (y, y) + (Ly, Ly) - \left[(Ly, y) + (y, L_y)\right]$$

so that

$$|Ju| \leq |u| \quad .$$

We can recover L from J by means of

$$(11) \qquad y = 1/2(Ju + u)$$
$$Ly = 1/2(Ju - u)$$

from which it follows that $I + J$ must be one-to-one and that D_L = range $(I + J)$. Note that J is closed if and only if range $(I - L)$ is closed and hence (by Remark 1) if and only if L is closed.

Conversely suppose that J is a contraction operator with $I + J$ one-to-one. Then (11) defines a dissipative operator L with

D_L = range $(I + J)$;in fact

$$(y, Ly) + (Ly, y) = 1/4 \left[(Ju - u,\ Ju + u) + (Ju + u,\ Ju - u) \right]$$
$$= 1/2 \left[|Ju|^2 - |u|^2 \right] \leqslant 0.$$

Finally we show that $I + J$ is automatically one-to-one if the range of $(I + J)$ is dense. For suppose there is a $u_0 \neq 0$ such that $Ju_0 + u_0 = 0$ and set $y = Jv + v$ for arbitrary $v \in D_J$. Then

$$|Jv + \alpha u_0|^2 = |J(v - \alpha u_0)|^2 \leqslant |v - \alpha u_0|^2$$

and expanding the extreme elements of this inequality gives

$$2 \operatorname{Re} \left[\alpha (u_0, y) \right] \leqslant |v|^2 - |Jv|^2.$$

Since this holds for arbitrary α we conclude that $(u_0, y) = 0$ and since y ranges over a dense set this implies that $u_0 = 0$, wnich is impossible.

Remark 3. It is always possible to extend a contraction operator J_0 to be a contraction operator with domain H . To accomplish this,first close up the operator and then set $Ju = 0$ for all $u \in D_{J_0}^{\perp}$.

We summarize the above in

Theorem 4. If L_0 is dissipative then J_0 defined as in (10) is a contraction and they are closed together. If J_0 is a contraction with range $(I + J_0)$ dense then L_0 defined as in (11) is dissipative with dense domain and conversely. The relations (10) and (11) establish a one-to-one inclusion preserving correspondence between dissipative extensions of L_0

(D_{L_0} dense) and contraction extensions of J_0. In particular, the maximal dissipative extensions of L_0 (D_{L_0} dense) correspond to contraction extensions of J_0 with domain H.

Corollary. If $\lambda > 0$ and L is dissipative with dense domain, then L is maximal dissipative if and only if range $(\lambda I - L) = H$.

Proof. L and $\lambda^{-1} L$ are dissipative and maximal dissipative together. But $\lambda^{-1}L$ is maximal dissipative if and only if

$$\text{range } (\lambda I - L) = \text{range } (I - \lambda^{-1}L) = H.$$

Proof of Theorem 3. If $\left[S(t)\right]$ is a semi-group of contraction operators then as we have seen above its generator A is dissipative with dense domain. Moreover by Theorem 2, $\lambda > 0$ belongs to the resolvent set for A and hence range $(\lambda I - A) = H$. Thus A is maximal dissipative. Conversely, if L is maximal dissipative with dense domain then for $\lambda > 0$, range $(\lambda I - L) = H$ by the Corollary to Theorem 4 and by Lemma 6, $|(\lambda I - L)^{-1}| \leqslant \lambda^{-1}$. Thus $R_\lambda(L)$ exists and is of norm $\leqslant \lambda^{-1}$. The Hille-Yosida Theorem applies and asserts that L generates a semi-group of contraction operators.

Lemma 7. If L is maximal dissipative with dense domain, then so is L^*, the adjoint of L.

Proof. Let J be the Cayley-transform of L. Then J is a contraction, $D_j = H$ and range $(I + J)$ is dense. Obviously J^* is a contraction with $D_{J^*} = H$. Suppose range $(I + J^*)$ were not dense, then $I + J$ would have a non-trivial zero which as we have seen in the proof of Theorem 4 is contrary to range $(I + J)$ being dense. As a consequence $I + J^*$ is one-to-one and defines a maximal dissipative operator, say , with dense domain. Now

$$(Ju - u,\ J^{*}v + v) = (Ju + u,\ J^{*}v - v)$$

so that

$$(Ly, z) = (y, Mz)\ , \qquad y \in D_L \quad , \quad z \in D_M \ .$$

This shows that $M \subset L^{*}$. On the other hand range $(I - M) = D_{J^{*}} = H$. Hence if L^{*} were a proper extension of M , then $I - L^{*}$ would have a non-trivial null vector, which is impossible when range $(I - L) = H$.

4- The double extension construction.

The previous extension theory has the shortcoming that the maximal dissipative operator L extending the dissipative operator L_0 not only has an unknown domain, but also has an unknown action on that part of D_L not in D_{L_0} . For most applications this is not good enough. Generally we are given both a 'minimal' operator L_0 and a 'maximal' operator L_1 and we wish to find a maximal dissipative operator L which is both an extension of L_0 and a restriction of L_1 . In this case the action of L is prescribed and only the domain of L is to be determined. In this direction we can prove

Theorem 5. If L_0 and M_0 are dissipative with dense domains as well as restrictions of each others' adjoints, then there exist maximal dissipative extensions $L \supset L_0$ and $M \supset M_0$ which are each others' adjoints.

Corollary. Under the hypothesis of this Theorem there exists a maximal dissipative operator L such that

$$L_o \subset L \subset M_o^* .$$

Example. Let $H = l_2$ and suppose (a_{ij}) is a matrix satisfying the hypothesis :

i) $\sum_j |a_{ij}|^2 < \infty$ for all i

$\sum_i |a_{ij}|^2 < \infty$ for all j

ii) each principal finite submatrix is dissipative.

Let $D_{L_o} = D_{M_o}$ = class of all ultimately zero vectors and set

$$(L_o y)_i = \sum_j a_{ij} \eta_j , \qquad y = \{\eta_j\} \in D_{L_o} ,$$

$$(M_o z)_i = \sum_j \bar{a}_{ji} \varphi_j , \qquad z = \{\varphi_j\} \in D_{M_o} .$$

It is clear that L_o and M_o are dissipative with dense domains and that they contain each others' adjoints. In fact

$$(M_o^* y)_i = \sum_j a_{ij} \eta_j$$

$$D_{M_o^*} = \left[y = \{\eta_j\} ; \quad \sum_i \Big| \sum_j a_{ij} \eta_j \Big|^2 < \infty \right] .$$

According to the Corollary to Theorem 3 there is a maximal dissipative operator L such that $L_o \subset L \subset M_o^*$ and L serves as the generator for the initial-value problem

$$\frac{d\eta_i}{dt} = \sum_j a_{ij} \eta_j , \qquad \{\eta_i(0) = c_i\} .$$

We will develop a solution to Theorem 5 couched in the language of subspaces of a Hilbert space. To this end we consider a Hilbert space H with inner product (x, y) on which there is given a continuous Hermitian symmetric bilinear form $Q(x, y) = (Wx, y)$, regular in the sense that W has a bounded inverse.

Definition. A subspace N (or P) will be called negative (positive) if

$$Q(x, x) \leq 0 \qquad (\geq 0)$$

for all $x \in N$ $(x \in P)$, and maximal negative (maximal positive) if N (or P) is not properly contained in any other negative (positive) subspace.

The ordered pair of subspaces $[N, P]$ will be called dual subspaces if N is negative, P is positive, and $Q(N, P) = 0$; and they will be called dual maximal subspaces if, in addition, N is maximal negative and P is maximal positive.

In therms of these concepts we shall prove

Theorem 6. Any pair of dual subspaces $[N_0, P_0]$ are contained in a pair of dual maximal subspaces $[N, P]$: $N \supset N_0$, $P \supset P_0$. Moreover N and P are complements of each other relative to Q.

That this implies Theorem 5 is seen as follows: Let $H = H_0 \times H_0$ with elements $x = [x', x^2]$, $y = [y', y^2]$, and inner product

$$(x, y) = (x', y') + (x^2, y^2) .$$

If we set

$$Q(x, y) = (x', y^2) + (x^2, y')$$

then the graph of L_0, $G(L_0)$, is a negative subspace and $G(-M_0)$ is a positive subspace whenever L_0 and M_0 are dissipative . Further if L_0 and M_0 contain each others adjoints then

$$Q([x', L_0x'], [y', -M_0y']) = -(x', M_0y') + (L_0x', y') = 0$$

and hence $[G(L_0), G(-M_0)]$ form a dual pair. Moreover Q is regular in this case

$$W[x', x^2] = [x^2, x'] \quad , \qquad W^2 = J \qquad , \quad \text{and} \quad W^{\times} = W.$$

Assuming Theorem 6 , there exists a maximal negative subspace $N \supset G(L_0)$ and a maximal positive subspace $P \supset G(-M_0)$ such that $Q(P, N) = 0$. If N and P are graphs, they are clearly graphs of maximal dissipative (respectively maximal accretive) operators. It suffices to consider the case of N .

Definition . If N is a subspace of $H = H_0 \times H_0$, its domain D_N is defined as

$$D_N = [x' ; [x', x^2] \subset N] \subset H_0 \quad .$$

Lemma 8. If N is a negative subspace with dense domain, then N is a graph.

Proof. If N were not a graph, then there would exist an $x = [x', x^2] \in N$ such that $x_1 = 0, x_2 \neq 0$. Since D_N is dense we can find a $y = [y', y^2] \in N$ such that $(y', x^2) \neq 0$. Then

$$Q(x, y) = (x'y^2) + (x^2, y') = 0 + (x^2, y') \neq 0 \; .$$

However on N , $-Q$ is a positive form and satisfies the Schwarz inequality, that is

$$|-Q(x,y)|^2 \leqslant [-Q(x,x)]\ [-Q(y,y)] = 0$$

since $Q(x,x) = 0$. Consequently N must be a graph.

Defining L and M so that $G(L) = N$, $G(-M) = P$ it is clear that L and M are each maximal dissipative operators. Since the graphs $G(L)$ and $G(-M)$ are dual it follows that $L \subset M^*$ and $M \subset L^*$. By Lemma 7 we know that L^* and M^* are (maximal) dissipative and hence $L = M^*$ and $M = L^*$. This concludes the proof of Theorem 5 from Theorem 6.

Before getting into the proof of Theorem 6 , it is convenient to re-norm H so that W becomes unitary. The original W splits H into its positive and negative complementary subspaces H_+ and H_- respectively and since W is regular there is an $m > 0$ such that

$$m^{-1}|x|^2 \leqslant Q(x,x) \leqslant m|x|^2$$

for all $x \in H_+$ or H_- . Further for any $x \in H$ there is a unique decomposition $x = x_+ + x_-$, $x_+ \in H_+$, $x_- \in H_-$ and

$$Q(x,y) = Q(x_+,y_+) + Q(x_-,y_-) \quad .$$

We now define

$$(x,y)_1 = Q(x_+,y_+) - Q(x_-,y_-)$$

and in view of the above inequality we get

$$1/m\,|x|^2 \leqslant |x|_1^2 < m\,|x|^2$$

Moreover if we set $W_1x = x$ for $x \in H_+$, $W_1x = -x$ for $x \in H_-$ we see that $Q(x, y) = (W_1 x, y)_1$. Since $W_1^2 = I$ this is the desidered normalization. From now on we drop the subscript 1.

Let E_+, E_- be the orthogonal projections into H_+ and H_- , respectively. For a given dual pair of subspaces $[N, P]$ with $x = x_+ + x_- \in N$ and $y = y_+ + y_- \in P$ we set

$$Jx_- = x_+ \quad , \quad Jy_+ = y_- \tag{12}$$

and extend J linearly on $E_-N + E_+P$. In the first place J is a well defined contraction operator. In fact, the inequalities

$$Q(x,x) = |x_+|^2 - |x_-|^2 \leqslant 0 \qquad \text{for } x \in N \text{ and}$$

$$Q(y, y) = |y_+|^2 - |y_-|^2 \geqslant 0 \qquad \text{for } y \in P \text{ imply}$$

$$|J(x_- + y_+)|^2 = |x_+ + y_-|^2 = |x_+|^2 + |y_-|^2 \leqslant |x_-|^2 + |y_+|^2 = |x_- + y_+|^2 .$$

Secondly, J is symmetric. To see this we note that for $x \in N, y \in P$ the duality of N and P gives

$$0 = Q(x, y) = (x_+, y_+) - (x_-, y_-) = (Jx_-, y_+) - (x_-, Jy_+) .$$

Hence for arbitrary $u, v \in D_y$

$$(Ju, v) = (Ju_- + Ju_+, v_- + v_+) = (Ju_-, v_+) + (Ju_+, v_-)$$

$$= (u_-, Jv_+) + (u_+, Jv_-) = (u, Jv) .$$

Thirdly, J anticommutes with W , that is $JW = -WJ$. In fact, $WD_J = D_J$ since $D_J = E_-N + E_+P$. Further

$$JWx = J(x_+ - x_-) = Jx_+ - Jx_- = -W(Jx_- + Jx_+) = -WJx .$$

Conversely given J a symmetric contraction which anti-commutes with W we see that $WD_J = D_J$ so that the same is true of $E_\pm = 2^{-1}(J \pm W)$. Setting $D_\pm = E_\pm D_J$ we have $D_\pm \subset D_J$, $D_J = D_+ + D_-$, and we may define the subspaces

$$(13) \qquad N = \left[x + Jx \; ; \; x \in D_-\right]$$
$$P = \left[x + Jx \; ; \; x \in D_+\right].$$

We shall show that N and P form a dual pair of subspaces. In the first place $JD_+ \subset H_-$ and $JD_- \subset H_+$. In fact if $x_- \in D_-$ and $y_- \in H_-$ then $Wy_- = -y_-$ so that

$$(Jx_-, y_-) = -(JWx_-, y_-) = (WJx_-, y_-) = (Jx_-, Wy_-) = -(Jx_-, y_-) .$$

Hence $Jx_- \perp H_-$, that is $Jx_- \in H_+$. Similarly $Jx_+ \in H_-$ for any $x_+ \in D_+$. It now follows from J being a contraction that N is a negative subspace and P is a positive subspace. To prove duality let $u \in N$, $v \in P$. Then $u_-, v_+ \in D_J$ and $u = u_- + Ju_-$, $v = v_+ + Jv_+$. Hence

$$Q(u, v) = (Ju_-, v_+) - (u_-, Jv_+) = 0$$

by the simmetry of J. Consequently $\left[N, P\right]$ form a dual pair.

It is clear that by starting with the dual pair $\left[N, P\right]$ and using (12) to define J, that we then return to the original dual pair by means of (13). Further an extending dual pair $\left[N_1, P_1\right]$ defines via (12) an extending operator J_1. Now N is a maximal negative subspace if and only if

$E_- N = H_-$ and P is a maximal positive subspace if and only if $E_+ P = H_+$. Hence if $D_J = H$ then the corresponding subspaces $[N, P]$ are dual maximal subspaces. We have therefore proved

Lemma 9. There is a one-to-one inclusion preserving correspondence given by (12) and (13) between dual pairs of subspaces and symmetric contractions which anti-commute with W. The dual pairs of maximal subspaces correspond in this way to symmetric contractions defined on all of H which anti-commute with W.

It remains to show that any symmetric contraction which anti-commutes with W can be extended to an operator with domain H having these same properties. M. Krein (Recucil Math. 20 (1947) pp 431-495) has shown how to extend a symmetric contraction J_0 to a symmetric contraction B defined in all H. Since $B \supset J_0$ and since $WJ_0 = -J_0 W$ we see for $x \in D_{J_0}$ that

$$WBW_x = WJ_0 W_x = -W^2 J_0 x = -J_0 x = -Bx \quad .$$

Hence $-WBW$ also extends J_0 as does

$$J = 2^{-1}(B - WBW) \; .$$

Thus so defined J is a symmetric contraction extending J_0 which anti-commutes with W_-. Consequently J defines via (13) a dual pair of maximal subspaces which extends the dual pair defining J_0.

Remark 1. If we set

$$\gamma_0 = \sup_{x \in D_{L_0}} \frac{(x, L_0 x) + (L_0 x, x)}{|x|^2 + |L_0 x|^2} \leqslant 0$$

then the previous extension can be made to a maximal dissipative operator L with corresponding $\gamma = \gamma_0$.

Remark 2. The above theorem can be applied to the study of conjugate symmetric dissipative operators and their extensions. To see what this means we recall that a conjugation C has the properties :

(i) $C(ax + by) = \bar{a}\, Cx + \bar{b}\, Cy$

(ii) $C^2 = I$,

(iii) $(Cx, Cy) = (y, x)$.

An operator L is called conjugate symmetric if $(Lx, Cy) = (Ly, Cx)$ for all $x, y \in D_L$, that is if $L \subset C L^{\times} C$. An operator L is called self-conjugate symmetric if $L = C L^{\times} C$. Using Theorem 6 it is not difficult to prove the following Theorem due to I. M. Glssman :

A dissipative conjugate symmetric operator with dense domain has a dissipative self-conjugate extension. We can dispense with the dense domain if $\gamma_0 < 0$. This problem is non-trivial even in the finite dimensional case. As a special case it solves the following : Given x_0, f such that

$$\gamma_0 = \frac{(x_0, f) + (f, x_0)}{|x|^2 + |f|^2} < 0 ,$$ to find a dissipative matrix operator

$L = (a_{ij})$ which is real symmetric $(a_{ij} = a_{ji})$ with $\gamma = \gamma_0$ and which takes x_0 into f_0 .

We conclude this section with a brief discussion of a duality theory for positive and negative subspaces.

Definition. Given a subset $S \subset H$, the Q-orthogonal complement S' of S is

$$S' \equiv \left[y;\ Q(x,y) = 0 \quad \text{for all} \quad x \in S \right] .$$

We denote the closure of S by $\bar{S}$.

It is clear that if $N(P)$ it is a negative (positive) subspace , then $\bar{N}(\bar{P})$ is also negative (positive) . Further S' is a closed subspace for any subset S . Likewise $S'' \supset S$ for any subset S and if $S_0 \subset S_1$ then $S'_0 \supset S'_1$.

Lemma 9. If S is a closed linear subspace then $S'' = S$.

<u>Proof.</u> From the fact $Q(x,y) = (Wx,y)$ we obtain

$$S' = (WS)^{\perp} , \tag{14}$$

and using the added fact that range $W = H$ we obtain

$$W S' = S^{\perp}$$

However if S is a closed subspace this implies $S = (WS')^{\perp}$. Applying (14) to S' instead of S gives $S'' = (W S')^{\perp}$, so that $S'' = S$.

<u>Lemma 10.</u> If N_0 and P_0 are Q-orthogonal complements with N_0 negative and P_0 positive, then $\left[N_0, P_0 \right]$ is a dual pair of maximal subspaces.

<u>Proof.</u> By Theorem 6 there exist $N \supset N_0$ and $P \supset P_0$ such that $\left[N, P \right]$ is a dual pair of maximal subspaces. However

$$N \subset P' \subset P'_0 = N_0 \qquad \text{and} \qquad P \subset N' \subset N'_0 = P_0$$

so that the maximality of N and P requires that $N = N_0$ and $P = P_0$

<u>Lemma 11.</u> If N is a maximal negative then $P = N'$ is maximal positive.

<u>Proof.</u> If N is maximal negative then N is closed and by Lemma

9, $N = N'' = P'$. In view of Lemma 10, it sufficies to show that P is positive. If this is not the case then there exists a $y \in P$ such that $Q(y, y)<0$. Let N_1 be the linear extension of N and y . Then for any $x \in N$

$$Q(x + ay, x + ay) = Q(x, x) + |a|^2 \; Q(y, y) \leqslant 0$$

so that N_1 is a negative subspace. Further $y \notin N$. For if $y \in N$ then since $y \in N'$ we get $Q(y, y) = 0$ contrary to one choice of y . It follows that N_1 is a proper extension of N which is negative and this is contrary to the fact that N is assumed to be maximal negative.

Lemma 12. A dual pair of maximal subspaces $[N, P]$ are necessarily Q - orthogonal complements.

Proof. By Lemma 11, N' is maximal positive and P' is maximal negative. By hypothesis $P \subset N'$, $N \subset P'$, and N, P are also maximal subspaces. It follows that $P = N'$ and $N = P'$.

5 - Boundary space theory.

One of the principal applications of dissipative operators is to the initial-value problem for partial differential equations where traditionally the problem of extending an operator has been put in terms of boundary conditions. As we shall see the extension problem takes on a somewhat simpler form in this case and is amenable to a more complete solution. A simple example will serve both as motivation and as a model for the development.

Let $H = L_2\,(0, 1)$

$$D_{L_o} = D_{M_o} = \left[\, y(x);\ y \text{ absolutely continuous};\ y, y' \in L_2\,(0, 1)\ ;\ y(0) = 0 = y(1) \,\right] ,$$

$L_o y = y' - y$,

$M_o z = -z' - z$

A simple calculation shows that

$$(L_0 y, y) + (y, L_0 y) = |y(1)|^2 - |y(0)|^2 - 2 \int_0^1 |y|^2 \, dx \leqslant 0$$

$$(M_0 z, z) + (z, M_0 z) = |z(0)|^2 - |z(1)|^2 - 2 \int_0^1 |z|^2 \, dx \leqslant 0$$

$$(L_0 y, z) - (y, M_0 z) = y(1)\,\overline{z(1)} - y(0)\,\overline{z(0)} = 0.$$

Thus L_o and M_o are both dissipative with dense domains and contained in each others' adjoints. The adjoints themselves can be shown to be

$$D_{L_o^{\times}} = D_{M_o^{\times}} = [\, y;\ y \text{ absolutely continuous; } y, y' \in L_2(0,1)$$

$$L_o^{\times} z = -z' - z \qquad \text{and} \quad M_o^{\times} y = y' - y.$$

We seek to characterize the maximal dissipative operators L such that $L_o \subset L \subset L_1 = M_o^{\times}$; in particular $D_{L_o} \subset D_L \subset D_{L_1}$. The problem can therefore be simplified by considering the factor space

$$\hat{H} = D_{L_1} / D_{L_o}$$

which is two dimensional space with coordinates $[a, b]$, where $y \in D_{L_1} \longrightarrow [a, b]$ means that $y(0) = a, y(1) = b$.

For $y \in D_L$, L dissipative

$$(Ly, y) + (y, Ly) = |y(1)|^2 - |y(0)|^2 - 2 \int_0^1 |y|^2 \, dx \leqslant 0 .$$

Actually this requires $|y(1)|^2 - |y(0)|^2 \leqslant 0$; for if $|y(1)|^2 - |y(0)|^2 > 0$ then since $D_{L_0} \subset D_L$ we can find a sequence $\{y_n\} \subset D_L$ such that

$y_n \longrightarrow [y(0), y(1)]$ and $\lim_n \int_0^1 |y_n|^2 dx = 0$. Hence

$$0 \geqslant (Ly_n, y_n) - (y_n, Ly_n) = |y(1)|^2 - |y(0)|^2 - 2\int_0^1 |y_n|^2 dx \rightarrow |y(1)|^2 - |y(0)|^2 > 0$$

as $n \longrightarrow \infty$,
which is impossible. Consequently $|y(1)|^2 - |y(0)|^2 \leqslant 0$ for all $y \in D_L$.

Thus relative to the form

$$\hat{Q}(\hat{y}, \hat{z}) = \hat{y}(1)\,\overline{\hat{z}(1)} - \hat{y}(0)\,\overline{\hat{z}(0)}$$

on $\hat{H}, D_L$ maps into a negative subspace $\hat{D}_L$. Such subspaces can only be zero or one dimensional. Since L is maximal dissipative $\hat{D}_L$ must be one dimensional, that is of the form $\hat{y}(1) = \alpha\,\hat{y}(0)$ when $|\alpha| \leqslant 1$. For every such choice of α we obtain a dissipative operator, say L, with domain of the form

$$D_L = \left[y;\ y \in D_{L_1}\quad ,\ \hat{y}(1) = \alpha\,\hat{y}(0)\right].$$

Each operator L of this type is maximal dissipative since by theorem 5 it would otherwise have a maximal dissipative extension L' such that $L \subset L' \subset L_1$ and this as we have just seen is impossible. It is also clear that these operators define all of the maximal dissipative operators between L_0 and L_1.

Thus the domain of all of the maximal dissipative operators between L_0

and L_1 are determined by and in one-to-one correspondence with the maximal negative subspaces of $\hat{H}$ relative to the form $\hat{Q}$. Now $\hat{Q}$ is included in $\hat{H}$ by the form

$$(15) \qquad Q(x,y) = (x', y^2) + (x^2, y') + 2(x', y')$$

on H, which for $x, y \in D_{L_1}$ becomes

$$Q([x', L_1x'], [y', L_1y']) = (x', Ly') + (L_1x', y') + 2(x', y')$$

$$= x'(1)\overline{y'(1)} - x'(0)\overline{y'(0)} .$$

This suggests that we use (15) as a starting point rather then the form Q used in the previous section. For $x \in G(L_0)$, $y \in G(L_1)$ we note that $Q(x,y) = 0$. Thus $G(L_0)$ is what we shall call a null space and $G(L_1) \subset G(L_0)'$. Actually $G(L_1) = G(L_0)'$ as can be verified. Finally we note that

$$\hat{H} = G(L_1)/G(L_0) .$$

This then is the pattern we shall follow in our present development.

We begin as before with a regular Hermitian bilinear form Q on H. A closed linear subspace S_0 is called a <u>null</u> space if $Q(x,x) = 0$ for all $x \in S_0$. It follows from the Schwarz inequality that $Q(x,y) = 0$ for all x, $y \in S_0$. Consequently if we set $S_1 = S_0'$ then $S_0 \subset S_1$. The quotient space

$$\hat{H} = S_1/S_0$$

will be called a <u>boundary space</u>.

In the usual topology for a quotient space, namely,

$$|\hat{x}| = \inf\left[|x|;\ x \in \hat{x}\right]$$

$\hat{H}$ is again a Hilbert space. In fact if β denotes the orthogonal projection of S_1 on $S_1 \ominus S_0$ then $|\hat{x}| = |\beta x|$. Moreover $\beta x = \beta y$ if and only if $x - y \in S_0$. Hence $\hat{H}$ is isomorphic and isometric with $S_1 \ominus S_0$. We shall also denote the natural map : $S_1 \to S_1 \ominus S_0$ by β and it will be clear from the context whether βy is to be thought of as lying in $\hat{H}$ or in $S_1 \ominus S_0$.

Lemma 13. If $x, y \in S_1$, then

(i)
$$Q(x, y) = Q(\beta x, \beta y)$$

(ii)
$$|Q(\beta x, \beta y)| \leqslant k\,|\beta x|\,|\beta y|$$

and if $Q(\beta x, \beta y) = 0$ for all $y \in S_1 \ominus S_0$, then $\beta x = 0$.

Proof. Now $Q(x, y) = 0$ for all $x \in S_0$, $y \in S_1$ and since $x - \beta x$ and $y - \beta y \in S_0$ it follows that

$$Q(x, y) = Q(\left[\beta x + (x - \beta x)\right],\ \left[\beta y + (y - \beta y)\right]) = Q(\beta x, \beta y).$$

This proves (i) and (ii) follows from the continuity of Q.
Finally if $Q(\beta x, \beta y) = 0$ for all $y = \beta y \in S_1 \ominus S_0$ then it holds for all $y \in S_1$ and by (i) $Q(\beta x, y) = 0$ for all $y \in S_1$. Thus $\beta x \in S_0 = S_1'$ as well as $S_1 \ominus S_0$ and so $\beta x = 0$.

As a consequence $Q(x, y)$ depends only on the cosets to which x, y belong. Thus Q induces a bilinear form $\hat{Q}$ on $\hat{H}$, namely

$$\hat{Q}(\beta x, \beta y) = Q(\beta x, \beta y).$$

Lemma 14. $\hat{Q}$ is a regular hermitian form on $\hat{H}$.

Proof. $\hat{Q}$ is clearly continuous and Hermitian so that $\hat{Q}$ defines a bounded self-adjoint operator $\hat{W}$ with the property

$$\hat{Q}(\hat{x}, \hat{y}) = (\hat{W}\hat{x}, \hat{y}) = Q(\beta x, \beta y) = (W\beta x, \beta y).$$

The last part of Lemma 13 also implies that $\hat{W}$ is one-to-one. It remains to show that $\hat{W}$ has a bounded inverse. To this end we let β_0 denote the projection of H on $S_1 \ominus S_0$. Then $\hat{W}$ can be thought of as a restriction of $\beta_0 W$ to $S_1 \ominus S_0$. Let $x \in S_1 \ominus S_0$. Then $W S_1 = S_0^{\perp}$ implies that $W^{-1} x \in S_1$. Further setting $W^{-1} x = y_0 + y_1$ when $y_0 \in S_0$ and $y_1 = \beta_0 W^{-1} x \in S_1 \ominus S_0$, we infer from the relation $W S_0 = S_1^{\perp}$ that $\beta_0 W y_0 = 0$ and hence that

$$\beta_0 W \beta_0 W^{-1} x = \beta_0 W y_1 = \beta_0 (x - W y_0) = \beta_0 x = x.$$

Hence $\beta_0 W^{-1}$ is a right inverse for $\beta_0 W$ on $S_1 \ominus S_0$ and this together with the fact that $\beta_0 W$ is one-to-one proves that $\hat{W}$ is regular.

Corollary. The previous duality theorem holds for the negative and positive subspaces of $\hat{H}$ relative to $\hat{Q}$.

Lemma 15. The mapping : $M \rightarrow \hat{M} = \beta M$ defines a one-to-one between the subspaces of S_1 which contain S_0 and the subspaces of $\hat{H}$. This correspondence preserves (a) negativity, (b) positivity, (c) inclusion, and (d) Q-orthogonal subspaces correspond to $\hat{Q}$-orthogonal subspaces. In particular, subspaces of S_1 which are maximal negative (maximal positive) relative to subspaces of S_1 contain S_0 and correspond to maximal negative (maximal positive) subspaces of $\hat{H}$.

Proof. Everything except the last assertion follows from the relation $Q(x,y) = Q(\beta x, \beta y)$ for $x, y \in S_1$. If N is maximal negative relative to subspaces in S_1, then since $Q(S_1, S_0) = 0$ we can adjoin S_0 to N without affecting its negativity. Consequently N must already contain S_0.

Lemma 16. If $\hat{M}$ is a subspace of $\hat{H}$ and $M = \beta^{-1}\hat{M}$, then $M' = \beta^{-1}\hat{M}' \subset S_1$.

Proof. Now $S_0 \subset M \subset S_1 = S_0'$ so that $S_1 = S_0' \supset M' \supset S_1' = S_0$. Thus $S_0 \subset M' \subset S_1$ and by Lemma 15, $\beta M'$ is orthogonal to $\beta M = \hat{M}$. As a consequence $M' \subset \beta^{-1}\hat{M}'$. Again by Lemma 15, $\beta^{-1}\hat{M}'$ is Q-orthogonal to M and hence $\beta^{-1}\hat{M}' \subset M'$. Thus $M' = \beta^{-1}\hat{M}'$.

Lemma 17. Suppose $\hat{N}$ is maximal negative in $\hat{H}$ and let $\hat{P} = \hat{N}'$. Then $\hat{P}$ is maximal positive, $N = \beta^{-1}\hat{N}$ is a negative subspace of S_1, $P = \beta^{-1}\hat{P}$ is a positive subspace of S_1, and both N and P are maximal with respect to the subspaces of H and Q-orthogonal complements.

Proof. It follows from Lemma 11 that $\hat{P}$ is maximal positive. That N is negative and P positive is clear from Lemma 15. By Lemma 16, $P = \beta^{-1}\hat{N}' = (\beta^{-1}\hat{N})' = N'$ and similarly $N = P'$. Finally Lemma 10, implies that N and P form a dual pair of maximal subspaces.

Corollary. Any negative (positive) subspace N of S_1 which is maximal negative (maximal positive) relative to subspaces of S_1 is also maximal negative (maximal positive) relative to the subspaces of H. Moreover $N' \subset S_1$ is maximal positive (maximal negative) relative to the subspaces of H.

We now have

Theorem 7. The mapping β^{-1} defines a one-to-one correspondence between the maximal negative (maximal positive) subspaces of $\hat{H}$ and the maximal negative (maximal positive) subspaces of H which lie in S_1. Under this correspondence $\hat{Q}$-orthogonal complements go into Q-orthogonal complements.

6- Operator theory.

We now apply the previous theory to operators in the following way. For H we take $H = H_o \times H_o$ with elements $[x', x^2]$, ... and inner product

$$(x, y) = (x', y') + (x^2, y^2)$$

and we take Q in the form

$$Q(x, y) = (x', y^2) + (x^2, y') - (Dx', y') = (Wx, y)$$

where D is a bounded negative self-adjoint operator. Parenthetically we note that the theory can be extended to the case where D merely negative self-adjoint.

In order to show that Q is regular we define the auxiliary operators on H :

$$U [y', y^2] = [y', -y^2 + Dy']$$

$$V [y', y^2] = [y^2, -y_1]$$

It is readily verified that $U^2 = I$, $V^2 = -I$, and $W = -V U$ so that W is regular. The operator U is of interest in its own right. In fact

(16) $$Q(Uy, Uz) = -Q(y, z)$$

(17) $$Q(y, Uz) = (y^2, z') - (y', z^2).$$

According to (16) the mapping $y \longrightarrow Uy$ defines a one-to-one inclusion preserving correspondence between the negative and positive subspaces of H which preserves Q-orthogonality. On the other hand (17) shows that if $N = G(L)$ where D_L is dense, then $U(N') = G(L^{\times})$.

We shall need the following Lemmas.

Lemma 18. If N is a negative subspace with dense domain then N is a graph.

Proof. Suppose N has an element of the form $y = [0, y^2]$, $y^2 \neq 0$; and choose $u = [u', u^2] \subset N$ so that $(u', y^2) \neq 0$; here we have used the fact that D_N is dense.

$$0 \geqslant Q(u + \alpha y, u + \alpha y) = Q(u, u) + 2 \operatorname{Re} \alpha (y^2, u')$$

for all α. This requires $(y^2, u') = 0$ which is impossible.

Lemma 19. If S is a subspace with dense domain, then S' is a graph.

Proof. If S' were not a graph it would contain an element of the form $y = [0, y^2]$, $y^2 \neq 0$. Choose $u = [u', u^2] \in S$ so that $(u', y^2) \neq 0$, again making use of the density of D_S. Then $u \in S$, $y \in S'$ implies

$$0 = Q(y, u) = (u', y^2),$$

which is impossible.

We come now to the main Theorem:

Theorem 8. Let S_0 be a null space with dense domain and set $S_1 = S_0'$ and

$$\hat{H} = S_1 / S_0 .$$

In this case S_0 and S_1 will be graphs of linear operators, say L_0 and

L_1 respectively. There is a one-to-one correspondence between the maximal negative subspaces $[\hat{N}]$ of $\hat{H}$ (taken relative to $\hat{Q}$) and the maximal dissipative operators $[L]$ such that $L_o \subset L \subset L_1$, this correspondence being defined by

(18) $$D_L = [y' ; \beta [y', L_1 y'] \in \hat{N}] ,$$

which is dense in H_o.

The adjoint transformation $M = L^{\times}$ is again maximal dissipative with dense domain and can be described as follows : $U(S_o)$ and $U(S_1)$ are also graphs of operators M_o and M_1 respectively. Let $\hat{P} = \hat{N}'$. Then $\hat{P}$ is maximal positive and $M = L^{\times}$ is such that $Mo \subset M \subset M_1$ and

(19) $$D_M = [z' ; \beta U [z', M_1 z'] \in \hat{P}] .$$

Proof. Since D_{S_o} is dense and S_o is a 'negative' subspace, Lemma 18 shows that S_o is a graph, say $G(L_o)$. On the other hand Lemma 19 asserts that S_1 is a graph, say $G(L_1)$. The space $\hat{H}$ is clearly a boundary space so that there exists a one-to-one correspondence between the maximal negative subspaces $[\hat{N}]$ of $\hat{H}$ and the maximal negative subspaces $[N]$ of H which are contained in S_1 , this correspondence is given by $N = \beta^{-1}(\hat{N})$. Since $N \supset S_o$ we see that D_N is dense and hence that N is a graph, say $G(L)$. Clearly $L_o \subset L \subset L_1$ and D_L is given by (18).

We now show that the operators L defined by the maximal negative subspaces contained in S_1 constitute all of the maximal dissipative operators between L_o and L_1 . In the first place if $G(L)$ is negative then L must be dissipative since

$$(Ly', y') + (y', Ly') = Q([y', Ly'], [y', Ly']) + (Dy', y') \leqslant 0$$

Conversely if L is a dissipative extension of L_o, then $G(L)$ is a negative subspace. In fact, given $y' \in D_L$ there exists a sequence $\{y'_n\} \subset D_L$ with $z'_n = y' - y'_n \to 0$ in H_o and $\{y'_n\} \subset D_{L_o}$. This is a consequence of D_{L_o} being dense and $y' + D_{L_o} \subset D_L$. By Lemma 13 (i)

$$Q([y', Ly'], [y', Ly']) = Q([z'_n, Lz'_n], [z'_n, Lz'_n]) \leqslant (Dz'_n, z'_n) \to 0,$$

so that $G(L)$ is a negative subspace. It follows that every maximal dissipative operator L extending L_o has a maximal negative subspace containing S_o for a graph and conversely. In particular the maximal dissipative operators L such that $L_o \subset L \subset L_1$ have graphs which constitute the maximal negative subspaces N such that $S_o \subset N \subset S_1$, which was to be proved.

As to the adjoint operators, we note that $D_{U(S_o)} = D_{S_o}$ and $U(S_1) = [U(S_o)]'$. Employing the above arguments we see that both the null space $U(S_o)$ and $U(S_1)$ are graphs, say of M_o and M_1 respectively. If $G(L)$ is maximal negative, then $G(L)' = \beta^{-1}(\hat{P})$ (hence $\hat{P} = \hat{N}'$) is maximal positive by Theorem 7 and so $U[G(L)']$ is maximal negative. Set $G(M) = U[G(L)']$, this will be a graph by Lemma 18. Clearly

$$U(S'_1) \subset U[G(L)'] = U[G(M)] \subset U(S'_o)$$

so that $M_o \subset M \subset M_1$. On the other hand

$$Q(U\left[G(M)\right],\ G(L)) = Q(G(L)',\ G(L)) = 0$$

from which it follows that $M = L^{\times}$ and hence is maximal dissipative with dense domain. Finnally $\beta\left[G(L)'\right] = \hat{P}$ is the same as $\beta\left\{U\left[G(M)\right]\right\} = \hat{P}$ which proves (19).

We now apply this theory to first order symmetric differential systems defined as follows:

Let G be a domain in R_n, $y(x)$ a vector-valued function defined on G to $\mathbb{C}_m$ with inner product $<,>$. We consider operators of the form

$$\sum_{i=1}^{n} A^i \frac{\partial y}{\partial x^i} + B\,y$$

where the $A^i(x)$ are continuously differentiable and symmetric matrix-valued functions in G, $B(x)$ is a continuous matrix-valued function in G and

$$D = B + B^{\times} + \sum_{i=1}^{n} \frac{\partial A^i}{\partial x^i} \leqslant \Theta\,, \qquad x \in G\,. \tag{20}$$

Note that G need not be bounded and that the growth of the coefficients is only mildly restricted; in particular the coefficients A^i can go to zero near the boundary Γ of G.

The Hilbert space H_0 is defined as the space of vector-valued functions on G with the inner product

$$(y', z') = \int_G < y'(x), z'(x) > dx\,.$$

We make one further assumption besides (20), namely

$$D \geqslant -K\,I\,, \qquad x \in G, \tag{21}$$

for some constant $K > 0$; In this case

$$\mathbb{D}: \qquad y \longrightarrow Dy$$

defines a bounded linear operator on H_0 . However this assumption is made only to simplify the exposition, an extended theory is available to handle the case where $\mathbb{D}$ is not bounded.

We define Q on $H = H_0 \times H_0$ by

$$Q(y, z) = (y', z^2) + (y^2, z') - (\mathbb{D}y', z') .$$

Set

$$D_{L_{00}} = [y'; y' \text{ continuously differentiable with compact support in } G]$$

$$L_{00}y' = \sum_{i=1}^{n} A^i \frac{\partial}{\partial x^i} y' + By'$$

For $y \in G(L_{00})$ we have

$$Q(y, y) = (y', \sum A^i y'_i + By') + (\sum A^i y'_i + By', y') - (\mathbb{D}y', y')$$

$$= \int_G \left[\sum < A^i y', y' >_i + < Dy', y' > \right] dx - (\mathbb{D}y', y')$$

$$= \int_\Gamma < A^n y', y' > dS = 0 \;;$$

here $A^n = \sum A^i n^i$ where n^i are the components of the outer normal to Γ. The last step is not necessarily meaningful since we have not assumed that Γ is smooth enough to have a normal. However since $y(x)$ has compact support in G , the last equality is readily verified.

Let S_o denote the closure of $G(L_{oo})$. It is clearly a null space with dense domain and hence the graph (Lemma 18) of some operator, say L_o . $S_1 = S_o'$ is again the graph of an operator, say L_1 . It can be shown that L_1 corresponds to a differential operator of the given type in some generalized sense. Defining the boundary space $\hat{H} = S_1/S_o$ we see that $\hat{H}$ is a space of cosets, each coset corresponding to a boundary data. A subspace of H is then a homogeneous boundary condition, a maximal negative subspaces of $\hat{H}$ correspond to the proper boundary condition, for the domain of the maximal dissipative operators L such that $L_o \subset L \subset L_1$.

References

1. Dunford and Schwartz, Linear operators, part I, Interscience Publ. (1958).
2. Friedrichs, Symmetric positive linear differential equations, Comm. Pure Appl. Math., Vol. 11 (1958) 333-418.
3. Hille-and Phillips, Functional analysis and semi-groups, Amer. Math. Soc. Coll. Publ. (1957).
4. Lax and Phillips, Local boundary conditions for dissipative symmetric linear differential operators, Comm. Pure Appl. Math., vol. 13 (1960) 427-455.
5. Lumer and Phillips, Dissipative operators in a Banach space, Pac. Jr. Math., vol. 11 (1961) 679-698.
6. Phillips, Dissipative operators and hyperbolic systems of partial differential equations, Trans. Amer. Math. Soc., vol. 90 (1959) 193-254.
7. Phillips, Dissipative operators and parabolic partial differential equations, Comm. Pure Appl. Math., vol. 12 (1959) 249-276.
8. Phillips, On the integration of the diffusion equation with boundary conditions, Trans. Amer. Math. Soc., vol. 98 (1961) 62-84.
9. Phillips, The extension of dual subspaces invariant under an algebra, Proc. Int. Symposium on Linear Spaces, Jerusalem (1960) 366-398.
10. F. Riesz and Sz. Nagy, Functional analysis, Ungar Publ., New York (1955).

CENTRO INTERNAZIONALE MATEMATICO ESTIVO

(C. I. M. E.)

LUIGI AMERIO

ALMOST-PERIODIC EQUATIONS IN HILBERT SPACES

ROMA - Istituto Matematico dell'Università

ALMOST-PERIODIC EQUATIONS IN HILBERT SPACES

by Luigi Amerio

1. - The present lecture concerns the existence of almost-periodic (a. p.) solutions of a. p. linear functional equations in Hilbert spaces.

We shall consider vibrating membrane variational equation and, afterwards, the general case of abstract almost-periodic equations (including partial differential equations whose coefficients depend on time t, besides space coordinates).

The main purpose is the extension to such equations of Bohr-Neugebauer's and Favard's theorems[(1)]. As it is well known, Bohr-Neugebauer's theorem, for ordinary differential equations with constant coefficients, asserts that if the known term is a. p., then every <u>bounded</u> solution is a. p.

Favard's theorems concern the general case, of ordinary differential systems whose coefficients and known terms depend almost-periodically on t. For these systems the boundedness of a solution does not imply, in general, its almost-periodicity : but it is possible to prove the existence of <u>one</u> a. p. solution, $\tilde{x}(t)$, by assuming the existence of a bounded solution $x_0(t)$: precisely $\tilde{x}(t)$ is characterized by the <u>minimax</u> property that the functional

$$\mu(x) = \sup_{t \in J} \|x(t)\| \qquad J = (-\infty < t < +\infty),$$

defined in the set of all bounded solutions, takes in $\tilde{x}(t)$ its minimum value.

It is essential, in the proof, to assume a suitable asymptotic behaviour for the bounded eigensolutions of homogeneous translated systems and their limits : the norms of such eigensolutions must have strictly positive infimums.

The circumstance that, for these problems, boundedness is related to almost-periodicity can be extended to some classical partial differential equations.

In this case the notion of almost-periodicity must be taken in the sense of the theory developed by Bochner [(2)] for vector valued functions, with values in a Banach space Y. The original theory, created by Bohr, was extended by Bochner to such functions and it is important, for what follows, to point a notable difference, concerning integration of a. p. functions, between numerical and abstract case.

Let f(t) be an a. p. function, from J to Y, and put

$$F(t) = \int_0^t f(t)\, dt \quad .$$

If Y is an euclidean space, then F(t) bounded $\Rightarrow$ F(t) a. p. (theorem of Bohr). For the general case (Y arbitrary Banach space) the same thesis was proved by Bochner under the more severe hypothesis that F(t) has a relatively compact range R_F. It was recently proved by Amerio [(3)] that, although Bochner's hypothesis cannot be substituted in the general case by Bohr's hypothesis, nevertheless this possibility exists if Y is a uniformly convex space : hence Bohr's theorem can be extended, in particular, to Hilbert spaces : Y Hilbert space, f(t) a. p. , F(t) bounded $\Rightarrow$ F(t) a. p.

This property, which belongs to the solutions of the very simple abstract differential equation $y' = f(t)$, gives rise, in a natural way, to a classification of Banach spaces, with respect to the connection between boundedness and almost-periodicity of solutions of differential equtions: this connection exists, as we shall see, for some problems interesting continuum mechanics.

2. - a) Let us consider the motion of a vibrating membrane Ω with fixed boundary, under an external force f, which depends on t, t $\in$ J, according an arbitrary law.

To describe the motion of such a membrane we have to solve the variational vibrating membrane equation (deduced from Hamilton's principle) :

$$(2.1)\qquad \int_J \{(x'(t),h'(t))_{L^2} - (x(t),\ h(t))_{H^1_0} + (f(t),h(t))_{L^2}\}\ dt = 0\ .$$

Equation (2. 1) is the weak form of the wave differential equation.

The notations are well known. Let us assume that Ω is a bounded, open and connceted set of the m-dimensional euclidean space R^m; then $(y_1, y_2)_{L^2}$, $(x_1, x_2)_{H^1_0}$ mean scalar products in L^2 and H^1_0 (Hilbert spaces); precisely : $L^2 = L^2(\Omega)$ is the space of all complex functions, square integrable in Ω ,

$$y = \{ y(\xi); \quad \xi = (\xi_1, \dots, \xi_m) \in \Omega \}$$

$$\|y\|_{L^2} = \left\{ \int_\Omega |y(\xi)|^2 d\Omega \right\}^{1/2};$$

$H^1_0 = H^1_0(\Omega) \subset L^2$ is the closure of $C^1_0(\Omega)$ (the set of all functions $x = \{x(\xi); \xi \in \Omega\}$, continuous with their first derivatives and with compact support) in the norm

$$\|x\|_{H^1_0} = \left\{ \int_\Omega \left(\sum_{j,k}^{1,\dots,m} a_{jk}(\xi) \frac{\partial x(\xi)}{\partial \xi_j} \frac{\partial \bar{x}(\xi)}{\partial \xi_k} + a(\xi)x(\xi)\bar{x}(\xi) \right) d\Omega \right\}^{1/2}.$$

Here $\bar{\alpha}$ means the conjugate of the complex number α ; moreover, the coefficients $a_{jk}(\xi) = \bar{a}_{kj}(\xi)$, $a(\xi) \geqslant 0$ are bounded and measurable in Ω , and it is

$$\sum_{j,k}^{1,\dots,m} a_{jk}(\xi)\lambda_j\bar{\lambda}_k \geqslant \delta \sum_1^m {}_k\ |\lambda_k|^2, \qquad \delta > 0\ .$$

In equation (2. 1), $x(t) = \{x(t,\xi); \xi \in \Omega\}$ means the unknown displacement of the membrane, $h(t) = \{h(t,\xi); \xi \in \Omega\}$ is a test function, $f(t) = \{f(t,\xi); \xi \in \Omega\}$ is the known external force. It is assumed that $x(t)$ and $h(t)$ are (strongly) continuous from J to H^1_0 , that their (strong) derivatives

$x'(t)$ and $h'(t)$ are continuous from J to L^2 and that $f(t)$ is locally Bochner-integrable ($\int_\Delta \|f(t)\|_{L^2} dt < +\infty$, $\forall$ compact Δ).

Moreover $h(t)$ has a compact support and equation (2. 1) must be satisfied for every test function $h(t)$.

One can prove that there exists, in J , one and only one solution $x = x(t)$ of (2. 1), satisfying the arbitrary initial conditions

$$x(0) = x_0 \in H_0^1 , \qquad x'(0) = x_1 \in L^2 .$$

Let us consider now the cartesian product $E = H_0^1 \times L^2$, whose elements are the pairs $X = \{x_0, x_1\}$, with $x_0 \in H_0^1$, $x_1 \in L^2$, and norms

$$\|X\|_E = \{ \|x_0\|^2_{H_0^1} + \|x_1\|^2_{L^2} \}^{1/2} .$$

E is a Hilbert space : the energy space.

Let $x = x(t)$ be a solution of (2. 1) : then $X(t) = \{x(t), x'(t)\}$ is a function from J to E (we shall still call $X(t)$ a solution) and it results

$$\|X(t)\|_E = \{ \|x(t)\|^2_{H_0^1} + \|x'(t)\|^2_{L^2} \}^{1/2} .$$

Hence $\|X(t)\|^2_E$ represents the total energy, at the time t , of the vibrating membrane Ω .

If $f(t) \equiv 0$ (that is for homogeneous wave equation) the principle of conservation of energy holds :

$$\|X(t)\|_E = \text{const.}$$

For the inhomogeneous equation, we can prove the following minimax theorem [(4)] (whose generalisation will be considered at § 4).

Assume that (2. 1) admits a bounded solution and let Γ_f be the set of all (necessarily bounded) solutions.

Put

$$\mu(x) = \sup_{t \in J} \| X(t) \|_E , \quad \tilde{\mu} = \inf_{X \in \Gamma_f} \mu(x)$$

there exists one and only one solution, $\tilde{x}(t)$, such that

$$\mu(\tilde{x}) = \tilde{\mu} .$$

We shall call $\tilde{x}(t)$ the minimal solution of (2. 1) : hence $\tilde{x}(t)$ is the solution which minimizes the supremum of the energy, in J .

b) The E-almost-periodicity of the solutions of the homogeneous wave equation ($f(t) \equiv 0$) has been proved, under more and more general hypotheses, by different authors : Muckenhaupt [5] (for $m = 1$), Bochner [6], Bochner and von Neumann [7], Sobolev [8], Ladyzenskaya [9]. Let us recall, what is interesting from physico-mathematical point of view, that Bochner has deduced from the principle of conservation of energy the almost-periodicity of a solution X(t), if its range R_X is relatively compact.

In other words, the solution X(t) has the following behaviour : taken an arbitrary $\varepsilon > 0$, there exists a relatively dense set $\{\tau\}_\varepsilon$ such that

$$\sup_{t \in J} \| X(t+\tau) - X(t) \|_E \leq \varepsilon , \qquad \forall \tau \in \{\tau\}_\varepsilon$$

Afterwards, Sobolev succeeded in eliminating compactness hypothesis : he proved in fact (under suitable smoothness conditions for the boundary of Ω) that every solution of the homogeneous equations is a. p. At last, the same thesis has been proved by Ladyzenskaya, without any smoothness condition.

c) Let us consider now the non-homogeneous wave equation, and assume that f(t) is L^2 - a. p. In this hypothesis, Zaidman [10] has extended Bochner's result, by proving that every solution X(t), which has a relatively compact range, is a. p. Afterwards, Amerio [11] has proved that every solution with bounded energy is a. p. : hence

$f(t)$ <u>a. p.</u>, R_X <u>bounded</u> $\Rightarrow X(t)$ <u>a. p.</u>

This statement, which constitutes an extension to the wave equation of Bohr-Neugebauer's theorem, can be proved by two different methods : the main tool, in the first method, is a suitable extension [12] to monotone sequences of real a. p. functions of Dini's theorem on monotone sequences of continuous functions, in a compact set. The second method [13] can be applied to much more general cases : one proves, first, that every bounded solution $X(t)$ is weakly a. p. (w. a. p.)(that is the scalar product $(X(t), G)_E$ is a. p., $\forall\ G \in E$). Afterwards, one proves that the range R_X is relatively compact.

It follows that $X(t)$ is (strongly) a. p., because of a general theorem on a. p. functions [14]

Contributions and generalisations, concerning the problem treated in [12], [13], have been subsequently given by Bochner [15], Zaidman [16], Prouse [17]. For C-almost-periodicity of solutions (in connection with recent results of Il'In [18] on regularisation of wave equation) see Vaghi[19].

3. - We expose now some recent results of Amerio[20], concerning the extension of Favard's theorems to linear second order a. p. functional equations in Hilbert spaces : in particular, to the weak form of hyperbolic equations, with coefficients depending on t (besides space coordinates). In agreement with general theories on these equations [21], we renounce to the continuity of the solutions $x(t)$, and we assume only that they are locally square-integrable.

Parallely, we are led to consider vector valued a. p. functions in the sense of Stepanov. In this connection, it is worth recalling Bochner's interpretation of Stepanov almost-periodicity (loc. cit. at (2)).

a) Let Y be a Hilbert space, $y = f(t)$ Y-a. p. -S^2, from J to Y : hence $f(t) \in L^2(\Delta, Y)$, $\forall$ compact Δ, and, taken an arbitrary $\varepsilon > 0$, there exists a relatively dense set $\{\tau\}_\varepsilon$, such that

$$\sup_{t \in J} \left\{ \int_{\Delta_0} \| f(t+\tau+\eta) - f(t+\eta) \|_Y^2 \, d\eta \right\}^{\frac{1}{2}} \leq \varepsilon \quad , \quad \Delta_0 = \left[-\frac{1}{2} \leq \eta \leq \frac{1}{2} \right] .$$

Let us consider now the Hilbert space $L_0^2(Y) = L^2(\Delta_0, Y)$; $g \in L_0^2(Y)$ means that

$$g = \{ g(\eta) ; \eta \in \Delta_0 \} \in L^2(\Delta_0, Y),$$

hence

$$\| g \|_{L_0^2(Y)} = \left\{ \int_{\Delta_0} \| g(\eta) \|_Y^2 \, d\eta \right\}^{\frac{1}{2}}$$

Put

$$\bar{f}(t) = \{ f(t+\eta) ; \eta \in \Delta_0 \} \quad ,$$

$\bar{f}(t)$ is a function from J to $L_0^2(Y)$ and it results

$$\| \bar{f}(t+\tau) - \bar{f}(t) \|_{L_0^2(Y)} = \left\{ \int_{\Delta_0} \| f(t+\tau+\eta) - f(t+\eta) \|_Y^2 \, d\eta \right\}^{\frac{1}{2}} .$$

Hence :

$$f(t) \in L^2(\Delta, Y), \forall \Delta , \Rightarrow \bar{f}(t) \ L_0^2(Y) \text{ - continuous in } J ,$$

$$f(t) Y\text{-a. p. -}S^2 \Leftrightarrow \bar{f}(t) \quad L_0^2(Y)\text{-a. p. (according Bochner's definition).}$$

Moreover, we shall say that $f(t)$ is Y-w. a. p. -S^2 if $\bar{f}(t)$ is $L_0^2(Y)$-w. a. p. , that is if, $\forall g \in L_0^2(Y)$, the scalar product

$$(\bar{f}(t), g)_{L_0^2(Y)} = \int_{\Delta_0} (f(t+\eta), g(\eta))_Y \, d\eta$$

is a Bohr a. p. function.

In what follows, we shall write $f(t)$ to indicate $\bar{f}(t)$, and we shall add the indication of the space where $f(t)$ has to be considered: thus the notations

$$f(t) \ Y\text{-a. p. -}S^2 \quad \text{or} \quad f(t) \ L_0^2(Y)\text{-a. p.} ,$$

$$f(t) \ Y\text{-w. a. p. -}S^2 \quad \text{or} \quad f(t) \ L_0^2(Y)\text{-w. a. p.} ,$$

are equivalent.

b) Let Y and $V \subseteq Y$ be two Hilbert spaces, V dense in Y : assume, moreover, that the immersion of V in Y is continuous ($\|x\|_Y \leq k\|x\|_V$, $k > 0$).

Let us consider the second order [22] linear functional equation :

$$
(3.1)\quad
\begin{aligned}
Q(x;h) = \int_J \Big\{ & (x'(t), h'(t))_Y - (A(t)x(t), h(t))_V + \\
& + (B(t)\, x'(t), h(t))_V + (C(t)x(t), h(t))_Y \Big\}\, dt = \int_J (f(t), h(t))_Y dt,
\end{aligned}
$$

where $x(t)$ (unknown function), $h(t)$ (test function), $f(t)$ (known term) satisfy, $\forall$ compact Δ, the conditions :

(3.2) $\quad x(t),\ h(t) \in L^2(\Delta, V)$; $\quad x'(t),\ h'(t) \in L^2(\Delta, Y)$;

$\quad h(t)$ has a compact support;

(3.3) $\quad f(t) \in L^2(\Delta, Y)$.

The derivatives are taken in the sense of distributions.

By (3. 2) and (3. 3), $x(t)$, $h(t)$ are $L^2_0(V)$ - continuous and $x'(t)$, $h'(t)$, $f(t)$ are $L^2_0(Y)$ - continuous.

Let us consider the following Hilbert space

$$W:\ w = \{w(\eta);\ \eta \in \Delta_0\}\ ,\ \{w(\eta)\} \in L^2_0(V),\ \ \{w'(\eta)\} \in L^2_0(Y),$$

with the scalar product

$$(w_1, w_2)_W = (w_1, w_2)_{L^2_0(V)} + (w_1', w_2')_{L^2_0(Y)}.$$

Hence, by (3. 2), $x(t) = \{x(t+\eta);\ \eta \in \Delta_0\}$ is a function from J to W, and it is

$$\|x(t)\|_W = \Big\{\int_{\Delta_0} \|x(t+\eta)\|^2_V d\eta + \int_{\Delta_0} \|x'(t+\eta)\|^2_Y d\eta\Big\}^{\frac{1}{2}}.$$

Therefore $x(t)$, $h(t)$ are W- continuous.

In equation (3. 1) the operators $A(t)$, $B(t)$, $C(t)$ are bounded, $\forall t \in J$; precisely :

$$A(t) \in \mathcal{L}(V,V), \quad B(t) \in \mathcal{L}(Y,V), \quad C(t) \in \mathcal{L}(V,Y)$$

and we assume, moreover, that $A(t)$, $B(t)$, $C(t)$ are continuous functions of t, in the uniform topologies of their spaces.

In particular, $A(t)$, $B(t)$, $C(t)$ can be supposed a. p. functions of t.

In this case we shall call regular any real sequence $\lambda = \{\lambda_n\}$ such that

(3. 4) $$\lim_{n\to\infty} A(t+\lambda_n) = A_\lambda(t), \quad \lim_{n\to\infty} B(t+\lambda_n) = B_\lambda(t),$$
$$\lim_{n\to\infty} C(t+\lambda_n) = C_\lambda(t)$$

uniformly on J. Hence $A_\lambda(t)$, $B_\lambda(t)$, $C_\lambda(t)$ are a. p. and, by fundamental Bochner's criterium, any real sequence contains a regular subsequence.

Moreover, we shall consider (according to Favard's theory for ordinary systems), $\forall$ regular λ, the homogeneous equation :

(3. 5) $$Q_\lambda(u;h) = 0 \quad , \quad Q_0(u;h) = Q(u;h) ,$$

that is the equation

$$\int_J \Big\{ (u'(t), h'(t))_Y - (A_\lambda(t)u(t), h(t))_V + (B_\lambda(t)u'(t), h(t))_V +$$
$$+ (C_\lambda(t)\, u(t)\, ,\, h(t))_Y \Big\} dt = 0 .$$

Let $z(t)$ be a W-bounded function. Put

$$\varphi(z;\tau) = \sup_{t \in J} \| z(t+\tau) - z(t) \|_W ,$$

let us call Λ_z the set of all W-bounded functions $x(t)$ such that

$$\varphi(x;\tau) \leq \varphi(z;\tau) \qquad \forall \tau \in J .$$

Let $\Lambda_{z,Q,f}$ be the set of all solutions $x(t)$ of (3. 1) such that $x(t) \in \Lambda_z$ and let $\Lambda_{z,Q}$ be the set of all eigensolutions $u(t)$ of the homogeneous equation

$$Q(u; h) = 0$$

such that $u(t) = x_2(t) - x_1(t)$, $x_i(t) \in \Lambda_{z,Q,f}$.

One can prove the following theorems

I (Minimax theorem) - Assume that :

α') there exists a W-bounded solution, $\bar{x}(t)$, of (3. 1);

β') $\forall\ u(t) \in \Lambda_{\bar{x},Q}$ it results

$$\operatorname*{Inf}_{t \in J} \| u(t) \| > 0 .$$

Then (3. 1) possesses, in $\Lambda_{\bar{x},Q,f}$, one and only one minimal solution, $\tilde{x}(t)$. Precisely, put

$$\mu(x) = \operatorname*{Sup}_{t \in J} \| x(t) \|_W ,$$

$$\tilde{\mu} = \operatorname*{Inf}_{\Lambda_{\bar{x},Q,f}} \mu(x)$$

there exists, in $\Lambda_{\bar{x},Q,f}$, one and only one solution, $\tilde{x}(t)$, such that

$$\mu(\tilde{x}) = \tilde{\mu} .$$

Let us observe, moreover, that if $A(t)$, $B(t)$, $C(t)$, $f(t)$ are periodic (with the same period) then $\tilde{x}(t)$ is periodic.

Let us assume now that : $A(t)$, $B(t)$, $C(t)$ are a. p. operators, $f(t)$ is $L^2_o(Y)$ - w. a. p. , $\lambda = \{\lambda_n\}$ is a regular sequence. Then there exists a suitable sequence (which we shall denote by $\{\lambda_n\}$) such that

$$\lim_{n\to\infty}{}^{*} f(t + \lambda_n) \underset{L^2_o(Y)}{=} f_\lambda(t) \qquad (f_\lambda(t)\ L^2_o(Y) \text{ - w. a. p. })$$

uniformly on J. Moreover, if condition α') holds, it is

$$\lim_{n\to\infty}{}^{*} \; \overline{x}(t + \lambda_n) \underset{w}{=} \overline{x}_\lambda (t),$$

where $\mu(\overline{x}_\lambda(t)) \leq \mu(\overline{x})$, $\varphi(\overline{x}_\lambda ; \tau) \leq \varphi(\overline{x}, \tau)$, and $\overline{x}_\lambda(t)$ satisfies the equation

$$Q_\lambda (x; h) = \int_J (f_\lambda (t), h(t))_Y \, dt .$$

Hence the set $\Lambda_{\overline{x}, Q_\lambda, f_\lambda}$ is not empty.

II (Weak almost periodicity theorem). Assume that :

α'') there exists a W-bounded solution, $\overline{x}(t)$, of (3. 1);

β'') $\forall$ regular λ and $\forall$ $u(t) \in \Lambda_{\overline{x}, Q_\lambda}$ it results

$$\operatorname*{Inf}_{t \in J} \|u(t)\| > 0 .$$

Then the minimal solution, $\tilde{x}(t)$, is W-w. a. p.

In other words, the scalar product

$$(\tilde{x}(t), g)_W = \int_{\Delta_0} \{ \tilde{x}(t + \eta), g(\eta))_V + (\tilde{x}'(t + \eta), g'(\eta))_Y \} \, d\eta$$

is a Bohr a. p. function, $\forall$ $g \in W$.

d) In order to prove that the minimal solution is W-a. p. , it is sufficient to prove that its range is W-relatively compact.

To that purpose, we shall restrict, first, the set where the minimal solution has to be defined.

Let us assume that there exists a W-bounded and W-uniformly continuous solution $\overline{x}(t)$.

Because of the inequality

$$\varphi(\tilde{x}; \tau) \leq \varphi(\overline{x}; \tau),$$

it follows

$$\lim_{\tau\to 0} \varphi(\tilde{x}; \tau) \leq \lim_{\tau \to 0} \varphi(\overline{x}; \tau) = 0 .$$

Hence the minimal solution $\tilde{x}(t)$ is W-uniformly continuous.

We enunciate now the conclusive statement.

III - (Almost-periodicity theorem). - Assume that :

α''') there exists a W-bounded and W-uniformly continuous solution, $\bar{x}(t)$, of (3. 1);

β''') $\forall$ regular λ and $\forall$ $u(t) \in \Lambda_{\bar{x}, Q_\lambda}$ it results

$$\operatorname*{Inf}_{t \in J} \| u(t) \| > 0 ;$$

γ''') the immersion of V in Y is completely continuous;

δ''') the hypothesis (of ellipticity) is satisfied :

$$(A(t)x, x)_V \geqslant \nu \|x\|_V^2 \qquad (\nu > 0) .$$

Then the minimal solution, $\tilde{x}(t)$, is W-a. p.

Let us observe, at last, that, for the problem of vibrating membrane, considered at § 2, conditions β'''), γ'''), δ''') are satisfied [(23)]

- - - - - - - - - - -

(1) Cfr. J. FAVARD, Leçons sur les fonctions presque-périodiques, Gauthier-Villars, Paris, 1933.

(2) S. BOCHNER, Abstrakte fastperiodische Funktionen, Acta Math., 61(1933).

(3) L. AMERIO, Sull'integrazione delle funzioni quasi-periodiche astratte, Ann. di Mat., 53(1961); Sull'integrazione delle funzioni quasi-periodiche a valori in uno spazio hilbertiano, Rend. Acc. Naz. dei Lincei, 28(1960). See also : M. L. RICCI, P. RIZZONELLI, Sulle funzioni l^1-quasi-periodiche, Rend. Ist. Lombardo, 95 (1961); L. AMERIO, Sull'integrazione delle

funzioni $l^p \{X_n\}$ -quasi-periodiche, con $1 \leq p < +\infty$, Ricerche di Mat., 12 (1963).

(4) L. AMERIO, Problema misto e quasi-periodicità per l'equazione delle onnon omogenea, Ann. di Mat., 49 (1960).

(5) C. F. MUCKENHAUPT, Almost-periodic functions and vibrating systems, Journ. of Math. and Phys., MIT, 8(1929).

(6) S. BOCHNER, Fast-periodische Lösungen der Wellen-Gleichung, Acta Math., 62 (1934).

(7) S. BOCHNER, J. VON NEUMANN, On compact solutions of operational differential equations, Ann. of Math., 36 (1935).

(8) S.SOBOLEV, Sur la presque périodicité des solutions de l'équations des ondes, I, II, III, Compt. rend. Ac. Sc. U. R. S. S. (1945).

(9) O. A. LADYZHENSKAYA, Mixed problems for hyperbolic equations, Moscov-Leningrad, 1953

(10) S. ZAIDMAN, Sur la presque-periodicité des solutions de l'équation des ondes non homogène, Journ. of Math. and Mech., 8 (1959).

(11) L. AMERIO, Quasi-periodicità degli integrali ad energia limitata dell'equazione delle onde con termine noto quasi-periodico, I, II, III, Rend. Acc. Naz. dei Lincei, 28 (1960).

(12) Cfr. (11), II. This theorem has been generalized to the almost automorphic functions : S. BOCHNER, Uniform convergence of monotone sequences of functions, Proc. Nat. Acad. Sci., U. S. A., 47 (1961).

(13) L. AMERIO, Sull'equazione delle onde con termine noto quasi-periodico, Rend. di Mat., 19 (1960).

(14) J. KOPEC, *On linear differential equations in Banach spaces*, Zeszyty Nauk Univ. Mickiewicza. Mat. Chem., I(1957); L. AMERIO, *Funzioni debolmente quasi-periodiche*, Rend. Sem. Mat. Univ. di Padova, 30 (1960).

(15) S. BOCHNER, *Almost-periodic solutions of the inhomogeneous wave equation*, Proc. Nat. Ac. Sc., 46(1960).

(16) S. ZAIDMAN, *Solutions presque-périodiques des équations hyperboliques*, Ann. Scient. Ec. Norm. Sup., 79(1962).

(17) G. PROUSE, *Analisi di alcuni classici problemi di propagazione*, Rend. Sem. Mat. Univ. di Padova, 32 (1962).

(18) V. A. IL'IN, *On solvability of mixed problem for hyperbolic and parabolic equations*, Uspehi Mat. Nauk, 15 (1960).

(19) C. VAGHI, *Soluzioni C-quasi-periodiche dell'equazione non omogenea delle onde*, Ricerche di Mat., 12(1963).

(20) L. AMERIO, *Sulle equazioni lineari quasi-periodiche negli spazi hilbertiani*, I, II, Rend. Acc. Naz. dei Lincei, 31 (1961); *Soluzioni quasi-periodiche delle equazioni lineari iperboliche quasi-periodiche*, Rend. Acc. Naz. dei Lincei, 33 (1962); *Soluzioni quasi-periodiche di equazioni quasi-periodiche negli spazi hilbertiani*, Ann. di Mat., 61(1963); *Su un teorema di minimax per le equazioni differenziali astratte*, Rend. Acc. Naz. dei Lincei, 1963; See also : L. AMERIO, *Sulle equazioni differenziali quasi periodiche astratte*, Ricerche di Mat., 9(1960), 10(1961); S. ZAIDMAN, *Solutions presque périodiques dans le problème de Cauchy, pour l'équation non homogène des ondes*, I, II, Rend. Acc. Naz. dei Lincei, 30(1961).

(21) Cfr. J. L. LIONS, *Equations différentielles-opérationelles et problèmes aux limites*, Springer, Berlin, 1961; *Equations différentielles-opérationelles dans les espaces de Hilbert*, course C. I. M. E.

(22) First order a. p. equation $x'(t) = Sx(t)+f(t)$ (S selfadjoint unbounded operator, f(t) a. p.) has been recently studied by ZAIDMAN (Teoremi di quasi-periodicità per alcune equazioni differenziali operazionali, Rend. Sem. Mat. e Fis. di Milano, 33 (1963)). Zaidman proves, first, that u'(t) is a. p. , by using spectral integral representation of the operator S; afterwards, by virtue of theorem on integration (loc. cit. at [(3)]) it follows that x(t) (bounded by hypothesis) is a. p.. Let us mention, moreover, on Navier-Stokes equation (first order, non linear, equation): C. FOJAS, Essais dans l'étude des solutions des équations de Navier-Stokes dans l'espace. L'unicité et la presque-periodicité des solutions "petites", Rend. Sem. Mat. Univ. di Padova, 32 (1962); G. PROUSE, Soluzioni quasi-periodiche dell'equazione di Navier-Stokes, Rend. Sem. Mat. Univ. di Padova (1963).

(23) Let us assume only that the immersion of V in Y is continuous: then (strong) almost-periodicity of $\tilde{x}(t)$ can be proved if equation (3, 1) admits a suitable theorem of continuous dependence (see L. AMERIO and S. ZAIDMAN, loc. cit. at [(20)]).

CENTRO INTERNAZIONALE MATEMATICO ESTIVO

(C. I. M. E.)

GIAN-CARLO ROTA

A LIMIT THEOREM FOR THE TIME-DEPENDENT EVOLUTION EQUATION

ROMA - Istituto Matematico dell'Università

A LIMIT THEOREM FOR THE TIME-DEPENDENT EVOLUTION EQUATION

by GIAN-CARLO ROTA

1. INTRODUCTION.

The topic of the present lecture is slightly at variance with those treated in the lecture series of this course. The main theme has been the abstract differential equation

(1) $$\frac{du}{dt} = A(t)u \quad , \qquad u \in X \quad , \qquad t \geqslant 0$$

where X is a Banach space, and where $A(t)$ is a family of unbounded linear operators.

The main problem has been the <u>solution</u> of (1) under very weak conditions on $A(t)$. Under suitable conditions, which we shall specify below, the solution $u(t)$ of (1) which for $t=0$ satisfies the initial condition $u(0) = f$, where $f \in X$, can be expressed in the form

(2) $$u(t) = P(t, 0)f \quad ,$$

where $P(t, s)$ is a family of <u>bounded</u> linear operators satisfying the <u>evolution equation</u>

(3) $$\begin{aligned} &P(t, s)P(s, r) = P(t, r), \\ &P(t, t) = 1. \end{aligned} \qquad t, s, r \geqslant 0$$

Thus, we shall understand the expression "solving the initial value problem for the equation (1)" as meaning that a family of bounded linear operators P can be determined which satisfies (3) and which gives a solution of (1) by formula (2).

Our present point of view will be to take the existence question of

(1) for granted, and to investigate instead another problem related to the evolution equation, which for many investigations is as important, if not actually move, as the existence-uniqueness question. This is the <u>limiting behavior</u> of solutions $u(t)$ as $t \longrightarrow \infty$. To get an idea of the kind of question we have in mind, let us consider first briefly the case where $A(t) \equiv A$ is independent of t. This case has been thoroughly studied : it is the theory of one-parameter semigroup of Hille and Phillips. The evolution operators (3) become in this case

$$P(t, s) = P^{t-s} ,$$

where the right side is given by a one-parameter semigroup P^t whose infinitesimal generator is the operator A. In this case it is well-known that the solution $u(t)$ of

$$(4) \qquad \frac{du}{dt} = Au \quad , \qquad u(0) = f \quad , \qquad t \geqslant 0$$

is given by $u(t) = P^t f$.

The behavior at infinity of $u(t)$ is described by the classical <u>ergodic theorem</u>. Under suitable assumptions on the semigroup P^t - which can be equivalently stated as conditions on the infinitesimal generator A - the limit

$$(5) \qquad \lim_{T \to \infty} \frac{1}{T} \int_0^T P^t f \, dt = \lim_{T \to \infty} \frac{1}{T} \int_0^T u(t) dt$$

exists in the norm of the given Banach space X. (Cfr. Dunford-Schwartz, Linear Operators, Vol. I, Ch. VIII).

The limiting relation (5) is of fundamental importance in many investigations, and expresses a property of the semigroup which is analogous to a "boundary behavior" of sorts . In fact, to quote Norbert Wiener, the ergodic theorem is a kind of fundamental theorem of calculus at infinity.

We shall be concerned below with the following two questions:

(a) When can the Cesàro limit in (5) be replaced by an ordinary limit?

(b) Is there a "natural" analog of the ergodic theorem for the general evolution equation, corresponding to the time-dependent differential equation (1)?

These questions of course have innumerable answers, which depend largely on the kind of convergence that is required (norm convergence, weak convergence etc.). The type of convergence we shall require below is dominated convergence in $L_p(S, \Sigma, \mu) = X$, $p > 1$. Specifically, we take $u(t)$ and f to belong to some Banach space $L_p(S, \Sigma, \mu)$ where $\mu(S) = 1$; the operators $A(t)$ and $P(t, s)$ will operate on these spaces only and will be subjected to the further conditions specified below.

We say that a sequence f_n in $L_p(S, \Sigma, \mu)$ converges dominatedly to f (in symbols, $f_n \underset{d}{\rightarrow} f$) when

(I) $f_n(s) \rightarrow f(s)$ for almost every s in S, that is, f_n converges to f pointwise; and

(II) the function $f^*(s) = \sup_n |f_n(s)|$ belongs to $L_p(S, \Sigma, \mu)$.

We hasten to add that from this point on all equalities will be understood in the "almost everywhere" sense. Thus, a function is defined almost everywhere, etc.

Our main concern will be to give an answer to questions (a) and (b) when convergence is understood in the dominated sense. Questions related to dominated convergence appear in many contexts in both analysis and probability theory, and are usually considerably more difficult than the analogous questions for norm convergence; we need only recall, beside the G. D. Birkhoff ergodic theorem, the Riesz-Calderon-Zygmund theory of Hilbert transforms and singular integrals, the martingale theorem (cf. below), the results relating to the almost-everywhere convergence of Fourier series and of more general orthogonal expansions (such as the Rademacher-Menchoff theorem), the various strong laws of large numbers, etc.

Dominated convergence is the most satisfactory type of convergence that can be hoped for in problems of both analysis and probability; it is the nearest to actual pointwise convergence. It is very desirable to have a unified theory relating to this kind of convergence, in much the same way as the theory of Banach spaces gives a unified approach to questions of convergence in the mean; unfortunately, no such theory exists at present, and we have to rely on various methods of varying degree of complication.

We can now proceed to state our main result. First, we consider the discrete analog of equation (1). This is the difference equation (for integer n)

$$(6) \qquad \Delta U_n = (P_n - I)U_n \ , \qquad n \geq 1.$$

The solution of this difference equation which satisfies the initial condition $u_1 = f$ is easily seen to exist uniquely when the P_n form a sequence of bounded operators in the Banach space X. Indeed, since $\Delta u_n = u_{n+1} - u_n$ we easily get

$$(7) \qquad U_{n+1} = P_n U_n = P_n P_{n-1} U_{n-1} = \dots = P_n P_{n-1} \dots P_1 f$$

The discrete analog of the evolution operators $P(t, s)$ of (3) are the operators

$$(8) \qquad P(n, k) = P_n P_{n-1} \dots P_k \ , \qquad n > k$$

$$P(n, n) = 1.$$

We now consider an adjoint equation to (6), (for fixed n), that is the equation

$$(9) \qquad \Delta V_k = (P^*_{n-k} - I)V_k \ , \qquad k \geq 1 ,$$

where P_k^* is the adjoint operator of P_k. The solution of this equation that at k = 1 takes a given value g is given by the following expression

(10) $$V_k = P^*_{n-k} P^*_{n-k+1} \cdots P^*_n g$$

We shall call (9) the adjoint equation of (6). The continuous analog of equation (9), related to (1), is the equation (for fixed t)

(11) $$\frac{dw}{dr} = A^*(t-r)w \, , \qquad r \leq t$$

where $A^*(r)$ is the adjoint of the operator $A(r)$. Luckily, here we can make the change of variable $t-r = s$, thereby getting

(12) $$\frac{dv}{ds} = -A^*(s)v \, , \qquad s \geq 0 \, .$$

We shall call (12) the adjoint equation to (1). The family of evolution operators which give the solution of (12), call them $P_{ad}(s,q)$, are easily obtained in terms of (3), as follows.

Consider the family of operators $P^*(t,s)$, $t > s$. Because $(AB)^* = B^* A^*$, the operators $P^*(t,s)$ do not satisfy the evolution equation (3), but

(13) $$P^*(s,r)\, P^*(t,s) = P^*(t,r), \qquad t \geq s \geq r \geq 0$$
$$P^*(s,s) = I \, .$$

We now set $P_{ad}(s,q) = P^*(q,s)$. It is then easy to verify that P_{ad}, thus defined is the family of evolution operators which solves (12).

We can now state the "natural" solutions of (12). The expression to be considered is the following :

(14) $$\lim_{t \to \infty} P_{ad}(t,0) P(t,0) f \quad .$$

We shall see that, under suitable general conditions, this limit will exist. We shall derive the existence of the limit in the sense of dominated convergence, but other similar theorems for other types of convergence are easily stated and proved; in a sense, the theorem we prove below is best possible (cf. Burkholder, Annals of Math. Stat., 1962).

2. ANALYTIC PRELIMINARIES.

Before we state our main theorem concerning the convergence of (14) and that of its discrete analog, we must review some of the analytic tools which will be essential steps in the proof. These are the following :

(α) Doubly stochastic operators. Let (S, Σ, μ) be a probability space, namely, a measure space of measure one. A doubly stochastic operator P is one that has the following properties: (1) $P1 = 1$, where 1 is the function identically equal to one; (2) $P^* 1 = 1$, where P^* is the adjoint of P ; (3) if $f \geq 0$ in $L_p(S, \Sigma, \mu)$, then $Pf \geq 0$, that is, P is a positive operator. Doubly stochastic operators are bounded operators of norm one in all $L_p(S, \Sigma, \mu)$, $1 \leq p \leq \infty$.

The theory of doubly stochastic operators is not well-developed except when (S, Σ, μ) is an atomic measure space with a finite number of atoms having equal measure. In this case we obtain the classical case of doubly stochastic matrices which are being thoroughly studied. More generally, if the atoms do not necessarily have equal measure, the notion of doubly stochastic operators and their study is related to the extensive theory of matrices with the unimodular property. It would be interesting to extend this theory to the case of a non-atomic measure space, in view of the difficulty of extending the Birkhoff-von Neumann theorem to this case.

Probabilistically, doubly stochastic operators are simply Markov processes with discrete parameters and with an invariant measure, for which the adjoint process also has an invariant measure. Such processes occur in the most disparate connexions.

(β) Conditional expectation. A projection E in $L_2(S, \Sigma, \mu)$ which is doubly stochastic is called a conditional expectation. There is an extensive theory and representation of such operators (cf. the texts of Doob and Loève, or the author's paper in Padova Rendiconti, 1959); however, we shall only need the following property : conditional expectation preserves dominated convergence. In other words if $f_n \xrightarrow[d]{} f$, then $Ef_n \xrightarrow[d]{} Ef$.

Roughly speaking, a conditional expectation has much the same properties as an integral, for example, the Lebesgue bounded convergence theorem holds.

(γ) Martingales. An increasing (decreasing) martingale E_n is a sequence of conditional expectations such that $E_n E_m = E_n$ for $n \geq m$ ($n \leq m$). There is an analogous notion of martingales with a conditions parameter. The theory of martingales has been developed largely by Doob. The main result is the martingale theorem, which states that if E_n is a martingale, then $E_n f \xrightarrow[d]{} f$ for f in L_p, $p > 1$ (actually the general theorem is more far-reaching).

The notion of a martingale is one of the most fundamental in analysis; there is no telling how vast their applications will be in future years, once analysts begin to realize their power. So far the notion has been used largely by probabilists.

(δ) Abstract L-spaces. An abstract L-space is a Banach lattice X such that if $f, g \geq 0$ in X, then $\|f+g\| = \|f\| + \|g\|$. For the theory of these spaces, see Day, Normed Linear Spaces.

The main result on abstract L-spaces we shall need is the following. Let X be a Banach lattice such that $\|1\| = 1$, where 1 is a lattice identity. (Cf. Birkhoff, Lattice theory). Then there exists a structurally unique probability space (S, Σ, μ) such that X is latticially isomorphic to $L_1(S, \Sigma, \mu)$. This theorem is due to Kakutani.

(ε) Infinitesimal doubly stochastic operators. In keeping with the spirit of this Symposium, we consider the conditions to be satisfied by the operator function $A(t)$ in (1) in order that the evolution operators (3) be doubly stochastic. These conditions were essentially determined by Phillips (Czech. Math. Journal, 1962), and we shall limit ourselves to stating them without proof, since we shall have no occasion of using them. They are: (1) $A(t)1 = 0$; (2) $A^*(t)1 = 0$; (3) for f in the domain of $A(t)$, and f in $L_p \cap L_q$; $(p^{-1} + q^{-1} = 1)$ we have $(A(t)f, f^+) \geq 0$, where $f^+ = \max(f, 0)$.

3. MAIN THEOREM.

After all these preliminaries we can now proceed to state our main result, in both discrete and continuous forms.

Theorem. (1) Let $P_1, P_2, \ldots, P_n, \ldots$ be an infinite sequence of doubly stochastic operators in $L_p(S, \Sigma, \mu)$, where $p > 1$ and where (S, Σ, μ) is a probability space. Then for f in $L_p(S, \Sigma, \mu)$

(15) $$\lim_{n \to \infty} P_1 P_2 \ldots P_n P_n^* P_{n-1}^* \ldots P_1^* f$$

exists in the sense of dominated almost everywhere convergence.

(2) Let $P(t, s)$ be a family of doubly stochastic evolution operators in $L_p(S, \Sigma, \mu)$, where again (S, Σ, μ) is a probability space (that is, operators satisfy (3)). Then, for f in $L_p(S, \Sigma, \mu)$ the limit

(16) $$\lim_{t \to \infty} P_{ad}(t, 0)P(0, t)f = \lim_{t \to \infty} P^*(0, t)P(0, t)f$$

exists in the sense of dominated almost everywhere convergence.

Proof. The main device in the proof consists in reducing the convergence result to an application of the martingale theorem. For simplicity we shall only prove the discrete case. The continuous case can be proved either by reducing it to the discrete case, or else by a similar construction to the one we shall use below for the discrete case. At any rate, the difficulties involved in passing from the discrete to the continuous case are purely technical and involve no new ideas.

We begin by constructing what we shall call the path space of the sequence $P_1, P_2, \ldots$ To this end, we consider an infinite product of replicas (S_n, Σ_n, μ_n) of (S, Σ, μ) for $n = 0, 1, 2, \ldots$ namely

$$(S', \Sigma', \mu') = \prod_{n=o}^{\infty} (S_n, \Sigma_n, \mu_n)$$

It is well known (cf. Halmos, Measure Theory) that this infinite product space is a well-defined probability space. We shall now consider an algebra $\mathcal{A}$ of real valued functions on (S', Σ', μ') defined as follows. A function F on (S', Σ', μ') is a function of infinitely many variables $F(s_o, s_1, s_2, \ldots)$, $s_i \in S_i$. We say that F belongs to $\mathcal{A}$ if F is a sum of finite products of the form

(17) $$f_o(s_o), f_1(s_1) \ldots f_h(s_h),$$

where $f_i \in L_\infty(S_i, \Sigma_i, \mu_i)$.

Clearly $\mathcal{A}$ is an algebra. We define a linear functional L on $\mathcal{A}$ as follows. We first define L on functions F of the form (17) by the formula

$$L(F) = \int_S f_o P_1 \left[f_1 P_2 \left[f_2 \ldots \left[P_h f_h \right] \right] \ldots \right] d\mu$$

and then extend L by linearity. It is easy to see, using the fact that $P_n 1 = 1$ for all n that L is well-defined.

We shall now verify that L has a very important property ; it is a <u>positive</u> linear functional. In other words, we shall prove that if F is in $\mathcal{A}$ and takes only non-negative values, then $L(F) \geq 0$. This appears at first sight not to be a trivial statement, because a function F in $\mathcal{A}$ is a linear combination of functions of the form (17), and each of the summands may take negative values, even though the sum is always non-negative. To prove this statement, we notice that if

$$F(s_o, s_1, s_2, \dots) = \sum_i f_o^{(i)}(s_o) f_1^{(i)}(s_1) \dots f_k^{(i)}(s_k) \geq 0$$

for all $s_o, s_1, \dots, s_k$, then, changing the variable s_k to s and remembering that P_k applies only to functions of the variable s, we get

$$\begin{aligned} P_k\Big(\sum_i f_o^{(i)}(s_o) f_1^{(i)}(s_1) \dots f_{k-1}^{(i)}(s_{k-1}) f_k^{(i)}(s)\Big) &= \\ = \sum_i f_o^{(i)}(s_o) f_1^{(i)}(s_1) \dots f_{k-1}^{(i)}(s_{k-1}) P_k f_k^{(i)}(s) & . \end{aligned} \tag{18}$$

This expression is non-negative for every value of the variables $s_o, s_1, \dots, s_{k-1}$, s, because P is a positive operator. Next, we change the variable s_{k-1} to s, and then apply the operator P_{k-1}, that is

$$\begin{aligned} P_{k-1}\Big(\sum_i f_o^{(i)}(s_o) f_1^{(i)}(s_1), \dots, f_{k-2}^{(i)}(s_{k-2}) f_{k-1}^{(i)}(s) P_k f_k^{(i)}(s) \Big) &= \\ \sum_i f_o^{(i)}(s_o) f_1^{(i)}(s_1) \dots f_{k-2}^{(i)}(s_{k-2}) P_{k-1}\Big[f_{k-1}^{(i)} P f_k^{(i)} \Big](s) & , \end{aligned} \tag{19}$$

and again this expression is positive, because P_{k-1} is a positive operator. Proceeding in this way down to $k = 1$, we finally obtain that

$$\sum f_o^{(i)} P_1\Big[f_1^{(i)} P_2 \Big[\dots \big[P_k f_k^{(i)} \big] \Big] \dots \Big](s) \geq 0$$

for all s in S. Hence, integrating, we get $L(F) \geq 0$ as we wanted to show. We now remark that the algebra $\mathcal{A}$ has a very important property.

If F belongs to $\mathcal{A}$, and $F^+ = \max(F, 0)$, then F^+ belongs to $\mathcal{A}$. This is clear if F is of the form (17) for then

$$F^+(s_0, s_1, \dots) = f_0^+(s_0)\, f_1^+(s_1) \dots f_k^+(s_k) .$$

Therefore, to establish the assertion, it suffices to consider the case when both F and G depend upon two coordinates only : for then an easy induction will yield the general statement. Let $F^- = -\min(F, 0)$. Then $F = F^+ - F^-$, and furthermore the functions F^+ and F^- have disjoint supports. From this, it is easy to see that if $F = P - Q$, where P and Q are non-negative functions with disjoint supports, then $P = F^+$ and $Q = F^-$. Using this fact, the statement we are to prove reduces to proving that F+G can be written as the difference of two non-negative functions with disjoint supports. Let

$$\begin{aligned} F &= \sum_i f_0^{(i)}(s_0)\, f_1^{(i)}(s_1) = \\ &= \sum_i \left[f_0^{(i)+}(s_0) - f_0^{(i)-}(s_0) \right] \left[f_1^{(i)+}(s_1) - f_1^{(i)-}(s_1) \right] = \\ &= \sum_i \left[f_0^{(i)+}(s_0) f_1^{(i)+}(s_1) + f_0^{(i)-}(s_0)\, f_1^{(i)-}(s_1) \right] - \\ &\quad - \sum_i \left[f_0^{(i)-}(s_0)\, f_1^{(i)+}(s_1) + f_0^{(i)+}(s_0)\, f_1^{(i)-}(s_1) \right] = \\ &= P - Q , \end{aligned}$$

say. The functions P and Q are non-negative and it is an easy verification that their supports are disjoint. Thus, $P = F^+$ and $Q = F^-$ as we wanted to show.

At this point we can define a <u>seminorm</u> on $\mathcal{A}$ as follows

$$\| F \| = L(F^+ + F^-) , \qquad L(1) = 1 . \tag{20}$$

Since L is a positive linear functional, the positivity and triangle inequality follow at once. Clearly, those G in $\mathcal{A}$ whose norm in 0 form a linear subspace N ; taking the quotient vector space $\mathcal{A}/N = B$ we obtain canonically a norm on B . We now complete B , thereby obtaining a Banach space C .

We now claim that C is an abstract L-space. This is intuitively clear, but for the sake of completeness let us verify the statement. First we claim that C is a Banach lattice. Now, $\mathcal{A}$ is certainly lattice-ordered; thus, to show that C is a Banach lattice, it suffices to show that B is a lattice, since the completion of a lattice-ordered vector space is again lattice-ordered, and is in fact a Banach lattice. Thus, it all boils down to showing that if $F \geq 0$ and $0 \leq G \leq F$, with $L(F) = 0$, then $L(G) = 0$. But this is an obvious consequence of the positivity of L , which gives $L(F) \geq L(G) \geq 0$.

The fact that $\| x+y\| = \| x\| + \| y\|$ for x and $y \geq 0$ in C is also an evident consequence of (20). Thus, C is an L-space.

We now apply Kakutani's theorem and represent C as $L_1(S_\infty, \Sigma_\infty, \mu_\infty)$, where $(S_\infty, \Sigma_\infty, \mu_\infty)$ is a probability space. This is the space in which we shall now define a martingale. We shall need the following well-known (Cf. e. g. Rota, Padova Rendiconti, 1959) lemmas.

Lemma 1. Let $(S_\infty, \Sigma_\infty, \mu_\infty)$ be a measure space, and let (a) Y be a closed subspace of $L_1(S_\infty, \Sigma_\infty, \mu_\infty)$ which is also closed under the lattice operations (that is a closed Banach sublattice of $L_1(S_\infty, \Sigma_\infty, \mu_\infty)$ (b) Z be a subspace of $L_\infty(S_\infty, \Sigma_\infty, \mu_\infty)$. Then

(a) There exists a σ-subfield Σ of the σ-field Σ_∞ such that

$$Y = L_1(S_\infty, \Sigma_\infty, \mu_\infty)$$

(b) There exists a σ-subfield Σ' of Σ_∞ such that the closure of Z in $L_1(S_\infty, \Sigma_\infty, \mu_\infty)$ is $L_1(S_\infty, \Sigma', \mu_\infty)$.

<u>Lemma 2.</u> Let T be a linear operator from $L_1(S, \Sigma, \mu)$ into $L_1(S_\infty, \Sigma_\infty, \mu_\infty)$ (both over probability spaces) such that $T1 = 1$ and $\|Tf\| = \|f\|$ for all $f \geq 0$. Then there is a σ-subfield Σ_o of Σ_∞ such that $L_1(S, \Sigma, \mu)$ is isomorphic to $L_1(S_\infty, \Sigma_o, \mu_\infty)$.

<u>Lemma 3.</u> Let Σ be a σ-subfield of Σ_∞, and let E be the orthogonal projection of $L_2(S_\infty, \Sigma_\infty, \mu_\infty)$ onto $L_2(S_\infty, \Sigma, \mu_\infty)$.

Then E is a conditional expectation. In particular, E can be extended to a bounded operator in $L_p(S_\infty, \Sigma_\infty, \mu_\infty)$, $1 \leq p \leq \infty$.

We shall make use of these lemmas presently. First, notice that there is a natural mapping of $L_1(S, \Sigma, \mu)$ into $L_1(S_\infty, \Sigma_\infty, \mu_\infty)$ given by $f(s) \to f(s_o) \to x$, where $x \in C$. This mapping preserves the norms of all positive functions. Hence, by Lemmas 2 and 1 (a), there exists a σ-subfield of Σ_∞ isomorphic to Σ : we shall identify such a σ-subfield with Σ altogether, and identify $L_1(S, \Sigma, \mu)$ with $L_1(S_\infty, \Sigma, \mu_\infty)$.

We now come to the crucial point. Consider the algebra $\mathcal{A}_n$ of all functions of $\mathcal{A}$ which are "independent" of the coordinates $s_o, s_1, \ldots, s_n$. Under the canonical map into B and C successively the closure of $\mathcal{A}_n$ in $L_1(S_\infty, \Sigma_\infty, \mu_\infty)$ is mapped into a subspace of the form $L_1(S_\infty, \Sigma_n, \mu_\infty)$, $n = 0, 1, 2, \ldots$, where $\Sigma_o = \Sigma$. Let E_n be the orthogonal projection of $L_2(S_\infty, \Sigma_\infty, \mu_\infty)$ into $L_2(S_\infty, \Sigma_n, \mu_\infty)$.

The crucial part of the proof consists in verifying that for f in $L_p(S_\infty, \Sigma, \mu_\infty)$, $E_o E_n f = P_1 P_2 \ldots P_n P_n^* P_{n-1}^* \ldots P_1^* f$. Once this is verified, the proof is completed by (a) remarking that E_n is a martingale, (b) applying the martingale convergence theorem; (c) remarking that the conditional expectation operator E_o preserves dominated convergence in L_p.

Let us first compute $E_n f$, remembering that f is in $L_p(S_\infty, \Sigma, \mu_\infty)$. Actually, it can be assumed that $f \in L_p \cap L_2$, since this subspace is dense. Let

$$P_n^* P_{n-1}^* \ldots P_1^* f(s_n) = g(s_n) \tag{21}$$

we shall prove that $E_n f(s_o, s_1, \ldots) = g(s_n)$. We have to show that g is the unique function such that

$$\int_{S_\infty} h\, f\, d\mu_\infty = \int_{S_\infty} h\, g\, d\mu_\infty \quad , \tag{22}$$

for all $h \in L_\infty(S_\infty, \Sigma'_n, \mu_\infty)$. But once (22) has been verified for h ranging over a dense subset of $L_\infty(S_\infty, \Sigma_n, \mu_\infty)$ it will be true for all h. We shall take this dense subset to be $\mathcal{Q}_n$ (or rather its image in L_∞, but for simplicity we are identifying the two). Thus, it suffices to choose $h(s_o, s_1, s_2, \ldots) = f_n(s_n) f_{n+1}(s_{n+1}) \ldots f_{n+k}(s_{n+k})$. Then the right side of (22) equals for $g = P_n^* P_{n-1}^* \ldots P_1^* f(s_n)$,

$$\int_S P_1 P_2 \ldots P_n \left[(P_n^* P_{n-1}^* \ldots P_1^* f) f_n P_{n+1} \left[f_{n+1} \ldots \right]\right] d\mu$$

Now, for a doubly stochastic operator, $\int Pq\,d\mu = \int (P^* 1) q\, d\mu = \int q\, d\mu$. Hence this last expression simplifies to

$$\int_S (P_n^* P_{n-1}^* \ldots P_1^* f) f_n P_{n+1} \left[f_{n+1} \ldots \right] d\mu =$$

$$= \int_S f P_1 P_2 \ldots P_n \left[f_n P_{n+1} \left[f_{n+1} \ldots \right]\right] \ldots \Big] d\mu =$$

$$= \int_{S_\infty} h\, f\, d\mu_\infty \quad ,$$

as we wanted to show. This shows that $E_n f = g$.

Next - and last - we prove that if $q = q(s_n)$, then

$E_0 q = P_1 P_2 \dots P_n q(s_0)$. This is very easy. It suffices to verify that for $f = f(s_0)$ we have

$$\int_{S_\infty} f P_1 \dots P_n q \, d\mu = \int_{S_\infty} f q \, d\mu_\infty$$

But the integral on the right is (by definition!)

$$\int_S f P_1 P_2 \dots P_n q \, d\mu \quad ,$$

hence the proof is complete.

4. APPLICATIONS.

It is easy to get now a solution of our problem (a). We have that $P^{2n}f$ converges dominatedly if P is selfadjoint. This last fact was discovered independently by E. M. Stein.

In the continuous case, we infer the dominated convergence of $P^t f$ as $t \to \infty$ if P^t, in addition to being doubly stochastic, is also selfadjoint.

The reader may perhaps wonder how the result of the main theorem (originally published in Bulletin of the AMS, 1962, p. 94) was arrived at. The answer is that doubly stochastic operators are related to conditional expectation operators in much the same way as contraction operators in Hilbert space are related to orthogonal projections. It is this analogy that has suggested the present result, although of course the proof has to be based upon entirely different principles.

As an interesting unsolved problem we shall mention the convergence of

$$(P_1 P_2 \dots P_k)^n f \quad ,$$

where P_i are doubly stochastic operators or even conditional expectations.

CENTRO INTERNAZIONALE MATEMATICO ESTIVO

(C. I. M. E.)

. S. ZAIDMAN

EXISTENCE AND ALMOST-PERIODICITY FOR SOME DIFFERENTIAL EQUATIONS IN HILBERT SPACES

ROMA - Istituto Matematico dell'Università

EXISTENCE AND ALMOST-PERIODICITY FOR SOME DIFFERENTIAL EQUATIONS IN HILBERT SPACES

by S. ZAIDMAN

§ 1. -

Let H be a Hilbert space; $(\, , \,)$ is the scalar product and $\| \quad \|$ the norm in this space.

Consider a linear closed operator A in H, with domain D_A dense in H, and let A^* be the adjoint of A.

Denote by J the real axis $-\infty < t < +\infty$, and by $K_A (K_{A^*})$, the class of functions $\varphi(t)$ defined for $t \in J$ with range in $D_A (D_{A^*})$, twice continuously differentiable in H, with $A\varphi$ $(A^*\varphi)$ continuous in H, and with compact support in J.

We say that the function $u(t) \in L^2_{loc}(J; H)$ (space of Bochner square integrable functions on every compact of J, with values in H), is a weak solution on J of the equation

$$u''(t) + Au(t) = f(t) \tag{1}$$

where $f(t)$ is given in $L^2_{loc}(J; H)$ if the relation

$$\int_J (u(t), \; \varphi''(t) + A^* \varphi(t))dt = \int_J (f, \varphi)dt \tag{2}$$

holds, for every $\varphi \in K_{A^*}$.

We give a sufficient condition on A (precisely on A^*), which ensures the existence of (at least) one solution of (2), when $f(t)$ is given in $L^2_{loc}(J; H)$ without any other assumption at the infinity.

This is not, naturally, a well-posed problem.

Definition. We say that A satisfies condition S if the family of operators $(A^* + \lambda^2)^{-1}$ depending of the complex parameter λ, is bounded in norm by M for all λ outside j intervals of length s, on a double sequence of vertical lines in the complex plane, $\{\mathrm{Re}\,\lambda = \sigma_n \to +\infty\}$ and $\{\mathrm{Re}\,\lambda = \sigma'_n \to -\infty\}$

Here j is an integer $\geqslant 0$, s a real $\geqslant 0$, and the location of the j-intervals may vary with the line.

(Comp. Definition, pag. 131, in Agmon-Nirenberg [1]).

We have the following

Theorem 1. If A is a linear closed operator in H with D_A dense in H, which satisfies condition S, then, for every $f(t) \in L^2_{loc}(J; H)$, there is at least a function $U(t) \in L^2_{loc}(J; H)$, solution of (2) for every $\varphi \in K_{A^*}$.

We prove this result in our paper [7] , by using an adaptation of a method given firstly by Malgrange in his paper on existence for general partial differential equations [6] . (See also the Ch. III in the book of Hörmander [5]).

In the proof of various lemmas we need also some tecniques given in Agmon-Nirenberg [1] .

One sees readily that every self-adjoint operator satisfies the condition S.

§ 2. -

After these general considerations, we restrict ourselves to some more special cases.

Firstly, when A is self-adjoint and $\geqslant \beta I$, β real, we can regularize the weak solution given in the Theorem 1, and obtain, in fact, the following

Theorem 2. *Let* A *be self-adjoint and* $\geq \beta I$, *and* f(t) *given* $\in L^2_{loc}(J; H)$. *If* $u(t) \in L^2_{loc}(J; H)$ *satisfies (2) for every* $\varphi \in K_A$, *then one has*

$$u'(t) \ , \quad u''(t) \in L^2_{loc}(J; H) \ , \quad u(t) \in D_A$$

almost every where in J, *and* $A\,u(t) \in L^2_{loc}(J; H)$.

Moreover, the relation (1) holds, almost-every where on $t \in J$.

This is proved (and also in a somewhat more precise, quantitative, version), in our paper [8], by using the tecniques of Agmon-Nirenberg ([1] - Ch. IV).

An immediate corollary of the Theorem 2, is that u(t) and u'(t) are continuous in H. This permits us to consider the problem of the bounded or almost-periodic solutions of (1), when f(t) is bounded or almost-periodic from $t \in J$ to H.

One proves easily (as indicated to us by Agmon), the

Theorem 3. *In the conditions of the Theorem 2, if* f(t) *is continuous and bounded from* $t \in J$ *to* H, *any bounded (from* $t \in J$ *to* H) *solution* u(t) *of (1), has also a bounded derivative.*

Now, consider f(t) almost-periodic in Bochner's sense [4] from $t \in J$ to H. We have the following

Theorem 4. *If* A *is self-adjoint and* ≥ 0, *and* f(t) *is an almost-periodic function from* $t \in J$ *to* H, *then any bounded solution (on* J) u(t) *of the equation (1), is almost-periodic with its derivative.*

This theorem is proved in [8] by using a method of orthogonal projections which reduces (essentially) our problem to the problem of integration of Hilbert space valued almost-periodic functions. Here there is an important theorem of Amerio [2, 3], which says that bounded integrals of almost-periodic functions in uniformly convex spaces are also almost-periodic.

The results expressed in the preceeding theorems generalize an early paper of the author on almost-periodicity for the Poisson equation [9] .

BIBLIOGRAPHY

[1] S. Agmon-L. Nirenberg, Properties of solutions of ordinary differential equations in Banach spaces, Comm. Pure Appl. Math., May 1963.

[2] L. Amerio, Sull'integrazione delle funzioni quasi-periodiche a valori in uno spazio hilbertiano, Rend. Acc. Naz. Lincei, 28, fasc. I, (1960).

[3] L. Amerio, Sull'integrazione delle funzioni quasi-periodiche astratte, Ann. di Matem. vol. LIII, 371-382 (1961).

[4] S. Bochner, Abstrakte Fast-periodische Funktionen, Acta Math., vol. 61, 149-184, 1933.

[5] L. Hörmander, Linear Partial Differential Operators, Springer-Verlag, 1963.

[6] B. Malgrange, Existence et approximation des solutions des equations aux dérivées partielles et des equations de convolution, Ann. Inst. Fourier, Grenoble, 271-355 (1955-56).

[7] S. Zaidman, Un teorema di esistenza globale per alcune equazioni differenziali astratte, Ricerche di Matem. (1964).

[8] S. Zaidman, Soluzioni quasi-periodiche per alcune equazioni differenziali in spazi hilbertiani, Ricerche di Matem. (1964).

[9] S. Zaidman, Quasi-periodicità per l'equazione di Poisson, Rend. Acc. Naz. Lincei, vol. 34, Marzo 1963.